TEACHER'S SUPPLEMENT

MATHEMATICS STANDARD LEVEL

FOR THE

INTERNATIONAL BACCALAUREATE

SECOND EDITION

Alan Wicks

Copyright © 2006 by Alan Wicks

All rights reserved. No part of this book shall be reproduced or transmitted in any form or by any means, electronic, mechanical, magnetic, photographic including photocopying, recording or by any information storage and retrieval system, without prior written permission of the publisher. No patent liability is assumed with respect to the use of the information contained herein. Although every precaution has been taken in the preparation of this book, the publisher and author assume no responsibility for errors or omissions. Neither is any liability assumed for damages resulting from the use of the information contained herein.

ISBN 0-7414-2155-0

Published by:

INFINITY
PUBLISHING.COM

1094 New DeHaven Street, Suite 100
West Conshohocken, PA 19428-2713
Info@buybooksontheweb.com
www.buybooksontheweb.com
Toll-free (877) BUY BOOK
Local Phone (610) 941-9999
Fax (610) 941-9959

∞

Printed in the United States of America
Printed on Recycled Paper
Published June 2006

Preface

This second edition *Teacher's Supplement* complements the textbook *Mathematics Standard Level for the International Baccalaureate, Second Edition*. The Supplement has two sections: a set of Internal Assessment Portfolio Assignments and a set of solutions to exercises in the companion textbook.

Some of the portfolio assignments in the first edition turned out to be unsuitable for the new criteria of assessment. In this second edition, fifteen new assignments replace the original ones; seven type I investigations and eight type II modeling assignments. They have been specifically designed so that there is reasonable scope for the use of technology in a creative way and also so that all achievement levels are available to the resourceful student.

In addition, an attempt has been made in the second edition to correct errors and omissions in the solutions. I am grateful to those people who kindly pointed out some of these errors.

Because this book contains resource material for teachers' use, those Internal Assessment Portfolio Assignments which carry the wording '© Alan Wicks 2004' may be photocopied. The normal requirement is waived here, and it is not necessary to write to Alan Wicks for permission. Teachers are encouraged to edit these assignments to suit their own particular needs. Complete details of the Internal Assessment requirements and advice for teachers are available in the *Mathematics SL Guide* and the *Teacher support material* provided by the International Baccalaureate and available via the International Baccalaureate's Online Curriculum Centre.

The second section of the *Supplement* consists of a complete set of solutions to the exercises given in the textbook. In most instances, complete solutions are provided. In a few instances, alternative solutions are also provided. The answers are given exactly or correct to three significant figures, as appropriate, unless a specific accuracy is stated in the question.

AW

CONTENTS

	Page
Table of Internal Assessment Portfolio Assignments	1
Guide to the Assignments	2
Guide to the Use of the Regression Model Features of the Graphing Calculator	6
Internal Assessment Assignments	9
Solutions to Textbook Exercises	
Unit 1	28
Unit 2	43
Unit 3	61
Unit 4	76
Unit 5	82
Unit 6	98
Unit 7	122

INTERNAL ASSESSMENT PORTFOLIO ASSIGNMENTS

The assignments in this section are intended for possible use in the completion of the Internal Assessment component of the IB Mathematics SL course. It is important to emphasize that any available sources may be used to find suitable Internal Assessment assignments, and, of course, teachers may write their own Internal Assessment assignments.

The ones shown in this section are offered for the convenience of the teacher and are intended to cover a wide range of topics. Many of these assignments have been tested on students taking the Mathematics SL course and have been edited as a result of unanticipated difficulties. In addition, they are designed to offer a varying level of challenge to students. Solutions to these assignments have not been provided since these assignments may be submitted to the IB as part of the students' final assessment for the course.

Following the examples from the *International Baccalaureate Mathematics SL Teachers support material*, many of the assignments require the use of regression and curve fitting. While this involves mathematics which is not specifically part of the content of the Mathematics SL syllabus it offers considerable scope for the use of technology. Although it is not advisable for teachers to involve their students in the finer points of regression analysis, a brief discussion with students about the use of the coefficient of correlation or the coefficient of determination is advisable. This will enable students to have a useful tool for deciding the most appropriate function to use for fitting a curve to a specific set of data. On page 6 is a guide to the use of the regression modeling features of the graphing calculator.

Table of Internal Assessment Assignments

Assignment	*Type*	*Title*	*Difficulty Level*
1	I	Intersections and Regions formed by Lines in a Plane	Straightforward
2	I	Tangents of a Parabola	Straightforward
3	I	How is the Shape of a Frequency Histogram Related to the Difference between the Mean and the Median of a Set of Data?	Straightforward
4	I	Summing the Terms of the Harmonic Series	Moderate
5	I	Matrix Investigation	Moderate
6	I	Regions cut off by Chords of a Circle	Challenging
7	I	Different Ways of taking Samples from a Population	Challenging
8	II	Growth of the Population of the United States	Straightforward
9	II	Packaging Tea	Straightforward
10	II	Can of Juice	Moderate
11	II	The Populations of Japan and Swaziland	Moderate
12	II	Crossing a River: a Vector Model	Moderate
13	II	Motion of a Ball in the Air	Moderate
14	II	Normal Potatoes	Moderate
15	II	Atmospheric Temperature and Pressure	Challenging

In order to assist teachers in their decision about which assignments to give their students the next section offers a guide to the Internal Assessment assignments. In this guide each assignment is given
- a difficulty level; straightforward, moderate or challenging
- a topic outline indicating the syllabus content of the assignment
- comments indicating what functions of the graphing calculator could be helpful in writing a solution to the assignment

Teachers should feel free to edit and revise these assignments to suit their own particular needs.

Guide to the Assignments

Assignment 1: Intersections and Regions formed by Lines in a Plane   **Type I**
Difficulty Level: Straightforward
Syllabus requirements:
- Obtaining the domain of a function
- Knowledge and recognition of a quadratic function

Comments: The graphing calculator can be used for regression modeling and curve fitting. The ability to draw large clear diagrams is essential in order that intersections and regions can be accurately counted.

Assignment 2: Tangents of a Parabola   **Type I**
Difficulty Level: Straightforward
Syllabus requirements:
- Drawing of a neat graph of a function and a tangent to a point on the graph
- Recognition of patterns of a sequence so that a general term for the sequence can be found
- Differentiation of simple polynomial functions

Comments: The graphing calculator can be used to construct tangents at a point on a curve. The graphing calculator can also be used for regression modeling and curve fitting although the patterns which occur are probably simple enough so that they can be recognized by direct observation.

Assignment 3: How is the shape of a Frequency Histogram related to the Difference between the Mean and the Median of a Set of Data?   **Type I**
Difficulty Level: Straightforward
Syllabus requirements:
- Knowledge of elementary statistics, including cumulative frequency curves and the ability to obtain the median from a cumulative frequency curve
- Understanding the meaning of a symmetric distribution and a skewed distribution

Comments: The graphing calculator can be used to carry out most of the operations involved in the statistics unit including entering data into a list, making a statistical plot, and obtaining the mean from grouped data.

Assignment 4: Summing the Terms of the Harmonic Series   **Type I**
Difficulty Level: Moderate
Syllabus requirements:
- recognition of patterns of a sequence so that a general term, expressed as a function of previous terms, can be found.

Comments: The graphing calculator can be used to obtain the minimum number of terms in a sequence whose sum is greater than a particular value. The graphing calculator can also be used to list terms and sum them.

Assignment 5: Matrix Investigation   Type I
Difficulty Level: Moderate
Syllabus requirements:
- Multiplication of matrices
- Recognition of patterns of a sequence so that a general term for the matrix can be found
- Knowledge of an infinite geometric series and its sum

Comments: Since the syllabus does not explicitly include powers of a matrix, teachers may want to discuss with students this minor extension of matrix multiplication. The syllabus only includes the concept of a limit in the introduction to differential calculus, therefore teachers may want to discuss limits with students in a broader context since in part 2 they are required to find the limit of a matrix as its power gets large. The graphing calculator can be used to multiply numerical matrices.

Assignment 6: Regions cut off by Chords of a Circle   Type I
Difficulty Level: Challenging
Syllabus requirements:
- Recognition of patterns of a sequence so that a general term for the sequence can be found
- Knowledge and recognition of a variety of functions
- Knowledge of Pascal's triangle

Comments: The graphing calculator can be used for regression modeling and curve fitting.

Assignment 7: Different Ways of taking Samples from a Population   Type I
Difficulty Level: Challenging
Syllabus requirements:
- Recognition of patterns of a sequence so that a general term for the sequence can be found
- Concept of taking random samples

Comments: The graphing calculator can be used for regression modeling and curve fitting. In addition, it may be useful to use the tabular facilities of the graphing calculator to express a function.

Assignment 8: Growth of the Population of the United States   Type II
Difficulty Level: Straightforward
Syllabus requirements:
- Knowledge of the properties of a variety of functions, particularly of exponential functions

Comments: The graphing calculator can be used to enter data into lists, carry out statistical plots and fit curves to a set of data points. Teachers may also want to discuss with students how to use search engines such as Google.

Assignment 9: Packaging Tea   Type II
Difficulty Level: Straightforward
Syllabus requirements:
- Use of algebra to obtain functions for the volumes of a variety of different shapes in terms of their linear parameters
- Knowledge of the formula for the volume of revolution of a solid

Comments: The graphing calculator can be used to find the maximum value of a variety of functions and also to find the area enclosed between the graph of a function and the *x*-axis. This assignment can easily be edited to include the opportunity to use differential calculus by requiring that students find maximum volumes and areas enclosed between the graph of functions and the *x*-axis by differentiation in parts (d), (g) and (h).

Assignment 10: Can of Juice   Type II
Difficulty Level: Moderate
Syllabus requirements:
- Use of algebra to obtain functions for the volume of a variety of different shapes in terms of their linear parameters
- Ability to draw large, neat diagrams

Comments: The graphing calculator can be used to find the maximum value of a function and also to enter data into a list, use a statistical plot, and fit a curve to a set of data points.

Assignment 11: The Populations of Japan and Swaziland   Type II
Difficulty Level: Moderate
Syllabus requirements:
- Knowledge of the properties of a variety of functions, particularly exponential functions

Comments: The graphing calculator can be used to enter data into lists, carry out statistical plots and fit curves to a set of data points. Teachers may also want to discuss with students how to use search engines such as Google.

Assignment 12: Crossing a River: a Vector Model   Type II
Difficulty Level: Moderate
Syllabus requirements:
- Knowledge of vector addition
- Ability to draw vector diagrams
- Ability to find the magnitude of a vector from its components
- Ability to use simple trigonometry to express the components of a vector

Comments: The graphing calculator can be used to enter data into lists, carry out statistical plots and fit curves to a set of data points.

Assignment 13: Motion of a Ball in the Air   Type II
Difficulty Level: Moderate
Syllabus requirements:
- Knowledge of kinematics.
- Ability to obtain a velocity function and a displacement function given an acceleration function

Comments: The graphing calculator can be used to enter data into lists, carry out statistical plots and fit curves to a set of data points.

Assignment 14: Normal Potatoes   Type II
Difficulty Level: Moderate
Syllabus requirements:
- Ability to find the mean and standard deviation of a sample
- Knowledge of the normal distribution
- Ability to draw a frequency histogram

Comments: The graphing calculator can be used to enter data into lists, form a frequency histogram and find means and standard deviations. Teachers may feel that the first table contains excessive data, in which case, they should feel free to delete some rows of data.

Assignment 15: Atmospheric Temperature and Pressure **Type II**
Difficulty Level: Challenging
Syllabus requirements:
- Concept of a function, with particular emphasis on piece-wise functions
- Good ability to carry out algebra involving logarithms

Comments: Use of the graphing calculator to enter data into lists, carry out a statistical plot, and fit a suitable curve to a set of data points.

A Guide to the Use of the Regression Model Features of the Graphing Calculator

Example 1 shows how to obtain the general term of a sequence by using the regression features of the graphing calculator.

Example 1: Find the general term of the following sequence of numbers: 5, 16, 39, 80, 145, 240, 371, 544, 765.

Solution 1: Enter the number of the term into one of the calculator lists and the sequence of numbers into another. Then go to the STAT CALC menu.

It is now necessary to choose from the menu of regression models the one which will provide a suitable fit. The accuracy of the fit can be measured by the coefficient of determination, R^2 (or r^2) and if $R^2 = 1$ then the regression model gives an exact fit to the sequence and the general term is obtained.

The next two screen displays show the results when a quadratic regression model is chosen.

```
QuadReg L1,L2
```

```
QuadReg
 y=ax²+bx+c
 a=15
 b=-59.2
 c=66
 R²=.9974786171
```

(If the coefficient of determination does not appear on the calculator screen it will be necessary to go to the catalog of your calculator and activate "DiagnosticsOn".)

Since $R^2 \neq 1$ the sequence of numbers does not fit a quadratic model and another regression model needs to be chosen. The next two screen displays show that a cubic regression equation gives an exact fit because $R^2 = 1$.

```
CubicReg L1,L2
```

```
CubicReg
 y=ax³+bx²+cx+d
 a=1
 b=0
 c=4
 d=0
 R²=1
```

Therefore $u_n = an^3 + bn^2 + cn + d$ where $a = 1$, $b = 0$, $c = 4$, $d = 0$ and the required general term is

$$u_n = n^3 + 4n$$

In fact, if a quartic regression model had been chosen, again $R^2 = 1$, showing that this model gives an exact fit but, since $a = 0$, the cubic nature of the sequence is confirmed, as shown, by the next two screen displays.

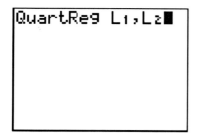

It is essential to use sufficient data in finding a regression model which fits the data. The next two screen displays show that if only three terms of the sequence are used a quadratic model will give a false exact fit, since a parabola can be drawn though any three points.

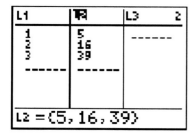

Example 2 shows how to fit the graph of a function to a set of data using the regression features of a graphing calculator.

Example 2: Find a function whose graph best fits the following set of data which shows the number of insects in a population at 10 day intervals.

Number of days	Number of Insects
0	6
10	12
20	30
30	60
40	135
50	279
60	630
70	1310
80	2830
90	6160

Solution 2: Since the data is statistical it is unreasonable to expect to be able to choose a regression model which gives a coefficient of determination, $R^2 = 1$. The aim, then, is to find a regression equation for which the coefficient of determination is as close to 1 as possible. The following screen displays show a number of choices of regression model and a decision can be made as to the best fit on the basis of the corresponding values of R^2 or r^2.

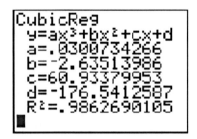
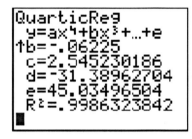

It is clear from these different regression curves that the one that provides the best fit is the exponential curve and the required function is $n = 5.98 \times 1.08^t$, where t is the time, in multiples of 10 days, and n is the number of insects. This might preferably be written in the form

$$n = 5.98 e^{\ln(1.08^t)} = 5.98 e^{t \ln 1.08} = 5.98 e^{0.0770t}$$

The following screen displays show how the data can be plotted and the curve fitted to the data.

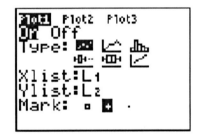
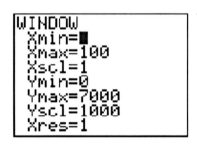
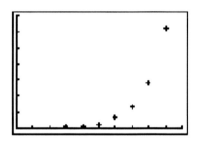

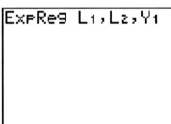
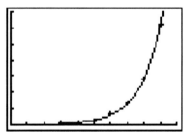

The regression model can be pasted directly into the function menu (see fourth screen display in the sequence) and the graph of the regression model can be superimposed on the data plot.

ASSIGNMENT 1   Type I

Title: Intersections and Regions formed by Lines in a Plane

The purpose of this assignment is to establish general formulas which describe the number of intersections and regions obtained when n lines are drawn in a plane.

Part 1:

(a) Draw diagrams which show how many intersections occur when n lines are drawn in a plane. Construct a table showing the number of intersections $I(n)$ for different values of n.

(b) Use your graphing calculator to find a quadratic function for $I(n)$ in terms of n.

(c) Check the validity of your function for more values of n.

(d) Discuss the scope and limitations of the domain of your function and also any other conditions which need to be imposed for the function to apply.

Part 2:

Carry out a similar investigation to find the number of regions $R(n)$ into which n lines divide a plane.

© Alan Wicks 2006

ASSIGNMENT 2   Type I

Title: Tangents of a Parabola

The purpose of this assignment is to investigate where tangents to points on the graph of a function of the form $y = x^n$ intersect the axes.

1. Draw a large neat graph of $y = x^2$ on a set of axes with $-5 \leq x \leq 5$ and $-5 \leq y \leq 5$. On it draw, as accurately as you can, the tangent at $x = 2$. Estimate the x and y intercepts of this tangent line.

2. Use your graphing calculator to construct the tangent to the graph of $y = x^2$ at $x = 2$ and find the x and y intercepts of this tangent line. Comment on any differences between these answers and those of part 1.

3. Find the x and y intercepts of the tangent to the graph of $y = x^2$ at $x = 1$ and $x = 3$. Suggest formulas for the x and y intercepts at $x = k$. In order to do this it may be necessary to obtain x and y intercepts of tangents at other points on the curve.

4. Carry out a similar investigation to find the x and y intercepts of the tangent to the graph of $y = x^3$ at suitable values of x and suggest formulas for the x and y intercepts at $x = k$.

5. Continue your investigation for tangents to the function $y = x^n$ for different values of n and suggest formulas for the x and y intercepts in terms of k.

6. Test the validity of your formulas with further examples and by finding the x and y intercepts of the tangent at $x = k$ to the graph of $y = x^n$. State any restrictions which may apply to values of n and k.

© Alan Wicks 2006

ASSIGNMENT 3

Type I

Title: How is the Shape of a Frequency Histogram Related to the Difference between the Mean and the Median of a Set of Data?

The purpose of this assignment is to investigate the difference between the mean and median of two different data sets: one obtained from a symmetric distribution of data and the other from a skewed distribution of data.

1. (a) Generate 30 random integers, X in the interval $1 \leq X \leq 25$.

    (b) Group the data into suitable class intervals and hence construct a frequency table of the data.

    (c) Draw a histogram of the data.

    (d) Calculate the mean value of X using the grouped data.

    (e) Construct a cumulative frequency table from the frequency table obtained in (b) and use it to draw a cumulative frequency curve.

    (f) (i) Use the cumulative frequency table obtained in (e) to estimate the median of X.

    (ii) If your graphing calculator claims to be able to calculate the median, use it to find the median and explain any discrepancy you notice between this answer and the one obtained in (f) (i).

2. Generate random numbers on your graphing calculator between 0 and 9 and count how many numbers, Y are generated before a '9' appears. Repeat this trial 30 times and then repeat, for the Y data, parts 1 (b) to 1(f).

3. (a) Compare the histogram for the X data with the histogram for the Y data.

    (b) Compare the difference between the mean and median of X and the mean and median of Y.

    (c) Make a conjecture about the relation between the shape of a histogram of a set of data and the difference between the mean and median of that data.

4. Test your conjecture by using examples of other data and discuss the scope and limitations of your conjecture.

© Alan Wicks 2006

ASSIGNMENT 4   Type I

Title: Summing the Terms of the Harmonic Series

The purpose of this assignment is to estimate the minimum number of terms of the harmonic series needed for the sum to exceed 10. The harmonic series is $1 + \frac{1}{2} + \frac{1}{3} + \frac{1}{4} + \ldots$

Part 1

(a) The purpose of part 1 is to direct your thinking in a way which may help you in part 2.

   Consider the sequence $\{a_n\} = \{1, 1, 2, 3, 5, 8, 13, \ldots\}$

(b) Write down the next three terms of the sequence and write the general term, a_n in terms of previous terms.

Part 2

The harmonic series is written out in such a way that the terms are grouped in brackets so that the number of terms in each grouping is the least required for their sum to be equal to, or greater than, 1.

The first three groupings are shown below.

$$(1) + \left(\frac{1}{2} + \frac{1}{3} + \frac{1}{4}\right) + \left(\frac{1}{5} + \frac{1}{6} + \ldots\right)$$

(b) Use your graphing calculator to form a series $u_1 + u_2 + u_3 + u_4 + u_5$, where u_i is the number of terms in the i th grouping. Note that $u_1 = 1, u_2 = 3$.

(c) Carry out an investigation in order to find a general term, u_n in terms of the previous terms. Now use your method of part (b) to find more values of u_n for higher values of n and check whether your general term agrees with these values.

(d) Use your general term to obtain an estimate of the minimum number of terms needed for the sum of the harmonic series to exceed 10.

(e) Discuss the accuracy of your result and explain why it is not exact.

© Alan Wicks 2006

ASSIGNMENT 5   Type I

Title: Matrix Investigation

Part 1: The purpose of part 1 is to investigate powers of matrices of the form $\begin{pmatrix} 1 & a \\ 0 & r \end{pmatrix}$

1. Let $\mathbf{A} = \begin{pmatrix} 1 & 2 \\ 0 & 3 \end{pmatrix}$. Find $\mathbf{A}^2, \mathbf{A}^3, \mathbf{A}^4$ and $\mathbf{A}^5$.

   Suggest a formula for $\mathbf{A}^n$, $n \in \mathbb{Z}^+$ in terms of n.
   Explain your reasoning for suggesting this formula.

2. Let $\mathbf{B} = \begin{pmatrix} 1 & a \\ 0 & r \end{pmatrix}$. Find $\mathbf{B}^2, \mathbf{B}^3$ and $\mathbf{B}^4$.

   Suggest a formula for $\mathbf{B}^n$, $n \in \mathbb{Z}^+$ in terms of n.
   Explain your reasoning for suggesting this formula.

3. Let $\mathbf{C} = \begin{pmatrix} 1 & 5 \\ 0 & 0.8 \end{pmatrix}$. Use your graphing calculator to find $\mathbf{C}^5, \mathbf{C}^{10}$ and $\mathbf{C}^{20}$.

   Suggest a limiting value for $\mathbf{C}^n$ as n goes to infinity.

Part 2: The purpose of part 2 is to investigate the powers of matrices whose rows have elements which add up to 1.

4. Let $\mathbf{P} = \begin{pmatrix} 0.3 & 0.7 \\ 0.1 & 0.9 \end{pmatrix}$. Find $\mathbf{P}^2, \mathbf{P}^3$ and $\mathbf{P}^4$ and investigate a value for $\mathbf{P}^n$ when n is very large.

   If the matrix $\mathbf{Q}$ is the limit of $\mathbf{P}^n$ as n goes to infinity try to write $\mathbf{Q}$ in an exact form.

5. Carry out an investigation on other 2×2 and 3×3 matrices with properties similar to $\mathbf{P}$ in part 4 to find if, when they are raised to large powers, similar results occur.

© Alan Wicks 2006

ASSIGNMENT 6 Type I

Title: Regions cut off by Chords of a Circle

The purpose of this assignment is to find the number of chords obtained, the number of intersections obtained, and the number of regions obtained, when n points positioned randomly on a circle, are connected by chords.

Consider a circle containing n randomly positioned points for $n \geq 2$. $C(n)$ is the number of distinct chords which can be drawn by joining two of the n points.

(a)   (i) Find $C(n)$ for various values of n displaying your results in a table.

         (ii) Use your graphing calculator to fit a quadratic function to the data in your table. Check the function you obtain using more data points.

Consider a circle containing n randomly positioned points for $n \geq 4$. $I(n)$ is the number of intersection points obtained when the n points are joined by chords.

(b)   (i) Find $I(n)$ for various values of n. Use Pascal's triangle to help you in obtaining more data points without the need to draw circles and count intersections. Display your results in a table, ensuring that you have at least 5 sets of data points.

         (ii) Fit a function of the form $I(n) = an^4 + bn^3 + cn^2 + dn + e$, $\quad a,b,c,d,e \in \mathbb{Z}$ to the data in your table.

Consider a circle containing n randomly positioned points. $R(n)$ is the number of regions which can be formed by drawing chords between the n points. The diagrams show the number of regions, $R(n)$ obtained when chords are drawn connecting the points for $n = 1, 2, 3$.

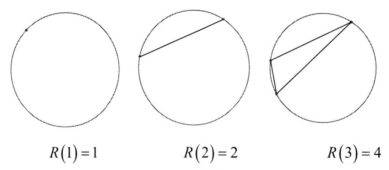

$R(1) = 1 \quad\quad R(2) = 2 \quad\quad R(3) = 4$

(c)   Draw similar diagrams in order to find $R(n)$ for other values of n.

(d)   Using your results from parts (a) and (b), or otherwise, obtain a formula for $R(n)$ in terms of n. State any limitations which may apply to this formula.

© Alan Wicks 2006

ASSIGNMENT 7   Type I

Title: Different Ways of taking Samples from a Population

The purpose of this assignment is to investigate the maximum number of different samples obtainable when a sample is taken from a population with replacement and when a sample is taken from a population without replacement.

Consider a bag containing n balls numbered from 1 to n and consider the following two ways of taking a random sample of 3 balls from the bag:

- A: Randomly select a ball, note its number and do not replace it. Do this trial three times. The sample consists of three different numbers in which the order of the numbers is unimportant. For example, different samples from a bag of 12 numbered balls could be $(1,8,11), (2,7,5), (6,2,1)$.

- B: Randomly select a ball, note its number and replace it. Do this trial three times. The sample consists of three numbers, not necessarily different, in which the order of the numbers is unimportant. For example, different samples from a bag of 12 numbered balls could be $(2,3,7), (8,4,5), (2,2,10), (5,5,5)$.

Part 1: Draw Pascal's triangle up to and including the 10th row.

Part 2:

a) (i) Use **method A** to find the number of different samples of size 3 which can be obtained from a bag containing n balls numbered from 1 to n for different values of n.

(ii) Use your results from (a) (i) and a graphing calculator to help you find a formula, in terms of n, for the number of different samples of size 3 which can be obtained from a bag containing n balls numbered from 1 to n.

b) (i) Use **method B** to find the number of different samples of size 3 which can be obtained from a bag containing n balls numbered from 1 to n for different values of n.

(ii) Use your results from (b) (i) and a graphing calculator to help you find a formula, in terms of n, for the number of different samples of size 3 which can be obtained from a bag containing n balls numbered from 1 to n.

Part 3:

Consider a random sample of 4 balls from the bag containing n numbered balls.

(c) Find formulas, in terms of n for the number of different samples of size 4

(i) using method *A* (ii) using method *B*

to take the random sample.

(d) Comment on the differences and similarities between the number of different samples obtained when the two sampling methods are used.

(e) If you wanted to take a sample of size 25 from a consignment of 1000 tomatoes in an attempt to estimate the mean weight of the tomatoes in the consignment which method of sampling would you choose? Give your reasons.

ASSIGNMENT 8   Type II

Title: Growth of the Population of the United States

The purpose of this assignment is to find functions which fit data for the United States population and to compare them with population projections obtained on the Internet.

The data in table 1 shows estimates of the population of the United States from 1900 to 1999. The data is provided by the US Census Bureau.

Year	Population (million)
1900	76.1
1910	92.4
1920	106
1930	123
1940	132
1950	152
1960	181
1970	205
1980	227
1990	249
1999	273

Table 1

(a) Draw a graph of the data shown in table 1 and find an exponential function relating the population to the time for the period from 1900 to 1999. Comment on how well your model fits the data.

(b) Use the exponential model you obtained in (a) to make a projection of the US population from 2000 to 2050.

(c) Find another model which gives a better fit to the population data and use it to modify your projection of the US population from 2000 to 2050.

(d) Find a website which makes population projections for the United States for the next 50 years and compare it with your projections.

© Alan Wicks 2006

ASSIGNMENT 9   Type II

Title: Packaging Tea

The purpose of this assignment is to investigate various shapes of box made with a fixed quantity of metal in order to find which shape gives the most volume.

Teakon makes metal boxes for packaging tea. The company uses 800cm^2 of metal for each box, which includes a base, four sides and a flat lid. Their traditional design has a rectangular base 10cm long and 8cm wide.

(a) Show that the height of the box is approximately 17.8 cm and find the volume of the box.

Thandile, one of the employees, suggests that a change in the dimensions, without using more metal, would give the box a larger volume. Suppose that the base of the box is rectangular with length twice the width.

(b) Find a formula for the height of the box in terms of the width of the base.

(c) Find a formula for the volume of the box in terms of the width of the base.

(d) Use your graphing calculator to find the maximum volume of the box and the dimensions of the box that give maximum volume.

(e) Is Thandile's suggestion correct? If so, how much more volume is obtained by the change in dimensions?

Another employee, Vusi, suggests that even more volume can be obtained using the same amount of metal by using a rectangular base in which the ratio of the length to the width of the base is $3:2$.

(f) Find the height and the volume of the box in terms of the width of the base. Draw a graph of the volume of the box against the width of the base.

(g) By investigating different lengths and widths for the base of the box, try to find a box whose surface area is 800cm^2 of metal which has a volume greater than 1510 cm^3.

A third employee, Sandile suggests yet another design. It is a container formed by rotating *ABDO*, shown in the diagram below, through $360°$ about the *x*-axis. *AB* is an arc of a circle centered at *C* and has the equation $y = \sqrt{69.3 - (x-5)^2}$, $0 \le x \le 10$. The base, along *AO*, and the lid, along *BD*, are circular with radius *OA* and *DB* respectively. The surface area of the this container is approximately 800 cm^2.

© Alan Wicks 2006

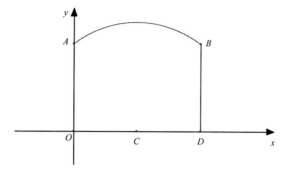

(h) Use your graphing calculator to find the volume of the container and discuss what other shapes might give even more volume for the same surface area of 800 cm².

ASSIGNMENT 10 Type II

Title: Can of Juice

Consider a can of juice of capacity 355 ml. The following information is given.

- The thickness of the curved surface of the can is 0.1 mm.
- The thickness of the base and lid are equal but unknown.

The main purpose of this assignment is to estimate the thickness of the base and the lid of the can.

1. Find a standard 355 ml juice can. Draw a diagram, or a number of diagrams, which show all the features of the can. [Do not cut open a can. The edges are very sharp!]

2. Use a perfect cylinder to model the can you drew in part 1 and define suitable variables for the radius and height of this cylinder.

3. By considering the volume of the perfect cylinder, establish a relation between the variables defined in part 2.

4. Assuming that the base and lid also have thickness of 0.1 mm, find the radius and height of the cylinder which minimizes the amount of aluminum used to make the can. (You will need to define a new variable for the volume of aluminum used.)

5. Find the radius and height of the can assuming that the base and lid have the following thicknesses (in millimeters) and that the amount of aluminum used is minimized in each case:
$$0.14,\ 0.18,\ 0.22,\ 0.26,\ 0.30$$
(Remember that the thickness of the curved part of the can is still 0.1 mm)
By comparing the dimensions of your model for these different thicknesses with the dimensions of the actual can, estimate the thickness of the base and lid. You may also assume that the actual can is designed to minimize the amount of aluminum used.

6. Discuss the accuracy to which it would be reasonable to give your result.

Consider a standard wine bottle of capacity 750 ml, which has an approximately cylindrical shape with radius 3.5 cm and height 20 cm above which it tapers to a short cylinder of radius 1.5 cm and height 10 cm.

7. Find whether a wine bottle with these dimensions minimizes the quantity of glass used to make it.

ASSIGNMENT 11 Type II

Title: The Populations of Japan and Swaziland

The purpose of this assignment is to develop a population model for a developed country (Japan) and for a developing country (Swaziland) and also to use them to make future population predictions for these countries.

The data in table 1 shows estimates of the population of Swaziland from 1911 to 2005. The data is taken from www.library.uu.nl/wesp/populstat/Africa/swazilac.htm

Year	Population (thousand)	Year	Population (thousand)
1911	100.0	1960	330.0
1921	112.8	1970	422.0
1927	122.0	1980	565.0
1936	156.7	1990	751.0
1944	171.3	2000	1083.3
1950	264.0	2005	1317.0

Table 1

(a) Find an exponential function which models the data from table 1 and comment on how well your model fits the data.

(b) Use the exponential function to predict the population for the years 2010, 2020 and 2030. Comment on the reliability of your prediction.

The data in table 2 shows estimates of the population of Japan for the period from 1900 to 2005

Year	Population (million)	Year	Population (million)
1900	43.8	1960	93.4
1910	49.6	1970	103.7
1920	56.0	1980	117.1
1930	64.5	1990	123.5
1940	71.9	2000	126.9
1950	83.2	2005	127.7

Table 2

(c) Find an exponential function which attempts to model the data from table 2. Comment on how well your model fits the data.

(d) Use the exponential function to predict the population for the years 2010, 2020 and 2030. Comment on the reliability of your predictions.

The Verhulst-Pearl Model may be a better model than the one obtained in (c). This model uses the function, $P(t) = \dfrac{MP_0}{P_0 + (M - P_0)e^{-kt}}$, where t is time, measured in years after 1900, and $P(t)$ is the population, measured in millions, at time t. k is approximately constant. P_0 is the population in 1900 and M is the limiting population (in millions) which may be considered to be the maximum population that the country could support under given social and environmental conditions. The limiting population is difficult to determine but in this assignment you should assume that $M = 150$.

(e) Use the data from table 2 and the Verhulst-Pearl Model to obtain a better model for the population of Japan and use it to predict the population of Japan in 2010, 2020 and 2030.

(f) Find a website which makes population projections for Swaziland and Japan for the next few years and compare it with your projections.

ASSIGNMENT 12
Type II

Title: Crossing a River: a Vector Model

The purpose of this assignment is to develop a vector model of a boat crossing a river.

A river 500 meters wide flows from west to east. P is a point on the south bank. Q on the north bank directly opposite P. Pedro has a boat that will travel at 1 ms^{-1} in still water. The current flows from west to east, and its speed varies from day to day.

On a Monday, the speed of the current is 0.75 ms^{-1}. Pedro wants to go in his boat from P to Q.

1. Define the necessary variables and draw a vector diagram which will enable you to calculate the time taken to go from P to Q.

On Tuesday the current has a speed of 2 ms^{-1}.

2. (a) Use vector diagrams to explain why Pedro cannot steer the boat directly from P to Q.

   (b) If Pedro points the boat in a direction 10° west of north and travels to the north bank reaching it at a point R, find the time taken for the boat to reach R and the distance QR.

3. Copy, complete and extend the table below by using the method of part 2(b), until you have sufficient data to plot a graph of the boat's direction against the distance QR.

Direction of boat west of north	Time to reach north bank	Distance of QR
10°		
30°		

   Use your graphing calculator to fit a suitable curve to the data points.

4. Use the function found in part 3 to estimate the direction Pedro should point the boat in order for QR to be minimized and find this minimum distance.

5. Test the accuracy of your answer to part 4 by using vectors and trigonometry to obtain a more accurate function relating the direction of the boat to the distance of QR. Comment on any discrepancies between your results in parts 4 and 5.

© Alan Wicks 2006

ASSIGNMENT 13   Type II

Title: Motion of a Ball in the Air

The purpose of this assignment is to investigate the motion of a ball thrown into the air.

A ball is thrown vertically upwards at a speed of 20ms^{-1}. The table shows the height of the ball at 0.4 second intervals. At the instant the ball is thrown $t = 0$.

Time	Distance
0	0
.4	6.9482
.8	11.965
1.2	15.236
1.6	16.874
2	16.931
2.4	15.427
2.8	12.409
3.2	7.694
3.6	2.2136

Part 1

    (a) Fit a suitable curve to the data in the table and use it to estimate
        (i) the maximum height reached by the ball and the time when that occurs
        (ii) the time taken from when the ball is thrown from ground level until it reaches the ground again.

Part 2

    (b) Consider an idealized situation in which the air resistance is neglected. Assuming that the ball is thrown vertically upwards with initial speed 20ms^{-1} and that the acceleration due to gravity is 9.81ms^{-2} downwards, find a function for the vertical height of the ball in terms of time.

    (c) Use this function to find
        (i) the maximum height to which the ball rises and when this occurs.
        (ii) the time taken from when the ball was thrown until it reaches the ground again.

Part 3

    (d) Compare your results from parts 1 and 2 and explain any differences.

    (e) If a ball is dropped from a balloon at a height of 100m estimate how much time elapses before it reaches the ground.

© Alan Wicks 2006

ASSIGNMENT 14

Type II

Title: Normal Potatoes

The main purpose of this assignment is to estimate the mean and standard deviation of a load of potatoes.

A company which produces potato chips obtains potatoes from a farmer. The company has agreed to purchase all potatoes whose weight is in excess of 250 kilograms.

The following table shows the weights of a sample of 150 potatoes taken from a delivery of potatoes to the company from the farm during one specific year.

258.9	288.5	266.3	267.6	252.0	282.5	262.1	288.2	263.6	287.8
264.5	252.9	266.5	269.9	267.0	275.1	255.2	257.4	273.4	264.5
260.7	256.3	273.4	265.8	261.5	272.7	262.0	273.2	250.8	272.7
267.0	267.7	285.0	253.8	275.0	279.0	263.6	289.2	275.9	290.9
267.3	261.8	267.3	308.8	265.8	273.0	271.4	250.8	252.3	256.0
278.8	270.7	260.3	265.2	253.1	252.9	267.1	265.9	258.4	258.4
257.3	258.2	250.4	268.9	250.8	289.6	308.1	275.4	267.2	314.9
267.1	280.0	286.1	267.5	271.1	269.4	272.0	266.7	315.5	265.3
259.5	253.5	303.0	285.0	291.6	265.1	268.0	270.2	270.1	277.4
262.0	267.3	277.7	275.1	280.3	255.5	261.3	274.5	255.9	274.9
260.0	251.3	256.5	278.1	289.2	258.7	275.8	256.8	300.7	255.8
278.9	254.0	281.0	250.9	238.8	319.9	268.0	292.3	260.9	271.7
266.0	251.9	256.1	253.3	267.0	271.7	280.3	271.1	269.5	254.8
259.8	254.3	252.7	269.5	259.1	275.3	276.8	317.3	250.2	260.1
273.9	264.7	287.9	260.2	271.4	270.6	270.6	310.7	283.4	298.8

(a) Group the data into suitable intervals and construct a frequency diagram.

(b) Draw a histogram of the grouped data.

- Assume that the weights of the potatoes from the potato crop grown by the farm are normally distributed.
- The farm states that the company has purchased 68% of its entire crop of potatoes.
- If X is normally distributed then $P(|X - \mu| < 3\sigma) \approx 0.995$

(c) (i) Draw diagrams which illustrate the bulleted information.

(ii) Estimate the mean and standard deviation of the farm crop of potatoes.

(iii) Comment on the accuracy of your answers.

© Alan Wicks 2006

(d) The following data is a sample of weights of 50 randomly selected potatoes delivered to the company from the farm the following year.

285.9	295.6	273.4	266.6	275.3	277.8	259.3	288.1	283.6	267.9
262.3	289.5	273.4	265.1	274.3	252.2	265.4	286.0	278.9	263.7
254.1	288.0	261.4	263.5	280.7	268.4	258.8	254.7	252.7	305.1
253.9	269.1	270.7	259.5	251.2	279.5	271.1	261.8	256.9	256.0
254.0	268.6	255.9	252.6	251.1	251.3	255.8	270.6	268.5	267.4

Use the data to determine whether the mean weight of potatoes from the farm is significantly different from the mean weight the previous year.

ASSIGNMENT 15   Type II

Title: Atmospheric Pressure and Temperature

The purpose of this assignment is to find graphically defined functions relating atmospheric temperature and pressure.

A meteorological laboratory has a remote controlled balloon which is being used to measure pressure and temperature at various altitudes from sea level to 20 km.

The data shown in the table below was obtained from an ascent of the balloon. Altitude is measured in kilometers above sea level, pressure is measured in atmospheres and temperature is measured in degrees Celsius.

Altitude	Pressure	Temperature
0	1	15.2
1	0.87360	10.1
5	0.50881	-10.8
10	0.25889	-35.0
15	0.13173	-49.8
20	0.06702	-43.9
25	0.03410	-40.0
30	0.01735	-34.5
35	0.00883	-29.5

Part 1

(a) Use the altitude and pressure data in the table to draw a graph relating these two variables.

(b) Find a suitable function which models this data. Comment on how well this function fits the data.

(c) Use the altitude and temperature data in the table to draw a graph relating these two variables.

(d) Find a suitable function which models this data. Comment on how well this function fits the data.

Part 2

The meteorologists at the laboratory want to find a function relating pressure to temperature.

(e) Use the data from the table to draw a graph relating pressure to temperature and find a suitable function which models this data.

(f) Using your results from Part 1 obtain a function relating pressure to temperature and compare it with the function you found in (e).

© Alan Wicks 2006

Solutions to Unit 1 Exercises

Exercise 1.1

1. (a) $\dfrac{1}{32}, \dfrac{1}{64}, \dfrac{1}{128}$ (b) $\dfrac{1}{6}, \dfrac{1}{7}, \dfrac{1}{8}$ (c) 13, 16, 19   (d) $-3, 3, -3$

    (e) 32, 64, 128   (f) 215, 342, 511   (g) 37, 60, 97   (h) 16, -17, 21

    (i) $\dfrac{11}{16}, \dfrac{13}{19}, \dfrac{15}{22}$ (j) 120, 720, 5040

2. (a) 2, 5, 8, 11   (b) -3, -1, 3, 11   (c) 1, 5, 11, 19   (d) $\dfrac{1}{2}, \dfrac{1}{3}, \dfrac{1}{4}, \dfrac{1}{5}$

    (e) 0, 2, 0, 2

3. (a) $1+2+3+4+5$ (b) $\dfrac{1}{2}+\dfrac{2}{3}+\dfrac{3}{4}+\dfrac{4}{5}+\dfrac{5}{6}$ (c) $0+\dfrac{1}{2}+\dfrac{2}{4}+\dfrac{3}{8}+\dfrac{4}{16}$ (d) $u_1+u_3+u_5+u_7$

    (e) $u_1^2+u_2^2+u_3^2+\ldots+u_n^2$ (f) $3+9+27+81+243$ (g) $1+1+1+1+1+1$

    (h) $1-2+3-4$

4. (a) $\displaystyle\sum_{r=1}^{5} r^{-1}$ (b) $\displaystyle\sum_{r=1}^{5} 2^{1-r}$ (c) $\displaystyle\sum_{r=1}^{5}(3r-5)$ (d) $\displaystyle\sum_{r=1}^{5}(r^3-1)$ (e) $\displaystyle\sum_{r=1}^{10}\sin\left(\dfrac{2r-1}{3}\right)\pi$

    (f) $\displaystyle\sum_{r=1}^{2n}(-1)^{r+1} u_r$

Exercise 1.2

1. (a) $a=1, d=2, u_5=1+4\times 2=9, u_{10}=19, u_n=2n-1$

    (b) $a=5, d=7, u_5=5+4\times 7=33, u_{10}=68, u_n=7n-2$

    (c) $a=\tfrac{2}{3}, d=\tfrac{5}{3}, u_5=\tfrac{2}{3}+4\times\tfrac{5}{3}=\tfrac{22}{3}, u_{10}=\tfrac{47}{3}, u_n=\tfrac{5}{3}n-1$

    (d) $a=4, d=-6, u_5=4+4\times-6=-20, u_{10}=-50, u_n=10-6n$

    (e) $a=-5k, d=4k, u_5=-5k+4(4k)=11k, u_{10}=31k, u_n=(4n-9)k$

2. The graphing calculator method described in Example 1.3 may be used to find n in each part of this question as an alternative to the method shown in part (a).

    (a) $u_n = 2n-1 = 99 \Rightarrow n = 50$, $S_{50} = 25(2+49\times 2) = 2500$ (b) $n = 27$, $S_{27} = 2592$

    (c) $n = 60$, $S_{60} = 2990$ (d) $n = 32$, $S_{32} = -2848$ (e) $n = 71$, $S_{71} = 9585k$

3. $a = 18$, $u_{23} = 656 \Rightarrow a + 22d = 656 \Rightarrow 22d + 18 = 656 \Rightarrow d = 29$

4. $a = 14$, $S_{13} = \frac{13}{2}(2\times 14 + 12d) = 572 \Rightarrow 13(14+6d) = 572 \Rightarrow d = 5$

5. $u_{18} = a + 17d = 469$, $u_{33} = a + 32d = 784$. Solving simultaneously gives $a = 112$ and $d = 21$.

6. $u_{10} = 25$, $u_{41} = 304 \Rightarrow a + 9d = 25$, $a + 40d = 304$. Solving simultaneously gives $a = -56$ and $d = 9$. $u_{75} = -56 + 74 \times 9 = 610$

7. (a) $u_n = 21 + (n-1)28 = 1197 \Rightarrow n = 43$ (b) $S_{43} = \frac{43}{2}(42 + 42 \times 28) = 26187$

8. $a = 110$, $d = -7$. We require the least value of n such that $a + (n-1)d < 0$. So, $110 + (n-1)(-7) < 0 \Rightarrow -7n + 117 < 0 \Rightarrow n > 16\frac{5}{7}$. Therefore the least value of n is 17, and the first negative term is $u_{17} = 110 + 16 \times -7 = -2$. The graphing calculator method described in Example 1.3 will provide a much quicker solution.

9. (a) $u_n > 2500$, $u_n = 1 + (n-1)14 = 14n - 13$ so that $14n - 13 > 2500 \Rightarrow n > 179.5$. Therefore the least value of $n = 180$.

    (b) $S_n = \frac{n}{2}(2 + 14(n-1)) > 2500 \Rightarrow 7n^2 - 6n - 2500 > 0 \Rightarrow n^2 - \frac{6}{7}n - \frac{2500}{7} > 0$

    $\Rightarrow \left(n - \frac{3}{7}\right)^2 - \frac{9 + 17500}{49} > 0 \Rightarrow n - \frac{3}{7} > 18.9 \Rightarrow n = 20$.

10. (a) $\sum_{r=1}^{n} r = 1 + 2 + \ldots + n$ so that $a = 1$ and $d = 1$. Therefore letting $\sum_{r=1}^{n} r = S_n$, $S_n = \frac{n}{2}(2a + (n-1)d) = \frac{n}{2}(2 + n - 1) = \frac{1}{2}n(n+1)$.

    (b) $S_n = 3n^2 + 5n$, $S_{n-1} = 3(n-1)^2 + 5(n-1) = 3n^2 - n - 2$. Now $S_{n-1} + u_n = S_n$
$\Rightarrow u_n = S_n - S_{n-1} = (3n^2 + 5n) - (3n^2 - n - 2) = 6n + 2$.
Therefore, $a = u_1 = 6 \times 1 + 2 = 8$, $u_2 = 6 \times 2 + 2 = 14$ so that $d = 6$.

11. (a) $d_u = 7$, $d_v = -5$

(b) $u_n = a_u + (n-1)d_u = 3 + (n-1)7 = 7n - 4$, $v_n = a_v + (-1)d_v = 115 + (n-1)(-5) = 120 - 5n$

(c) If $u_n > v_n$ then $7n - 4 > 120 - 5n \Rightarrow n > 10\frac{1}{3}$, so the least value of n is 11.

12. $u_k = 594 + (k-1)(-4) = 598 - 4k$ and $S_k = \frac{k}{2}(1188 + (k-1)(-4)) = 596k - 2k^2$.
As $u_k = S_k$, $598 - 4k = 596k - 2k^2 \Rightarrow k^2 - 300k + 299 = 0 \Rightarrow (k-299)(k-1) = 0$; therefore, $k = 1$, $k = 299$ but $k \neq 1$, so $k = 299$.

Exercise 1.3

1. (a) $a = 10\,000$, $d = 1\,200$. Note that u_2 is the value of the investment after 1 year. Therefore u_6 is the value of the investment after 5 years.

    (b) (i) $u_6 = a + 5d = 10\,000 + 6\,000 = 16\,000$ (ii) $u_{11} = 22\,000$ (iii) $u_{26} = 40\,000$

2. (a) $a = 8$, $d = -3$. Note that the 3rd term of the sequence represents the velocity after 2 seconds and the 4th term of the sequence represents the velocity after 3 seconds and so on.

    (b) (i) $u_4 = a + 3d = 8 - 9 = -1$, so velocity after 3 seconds is -1ms^{-1}.
    (ii) $u_{21} = a + 20d = 8 - 60 = -52$, so velocity after 20 seconds is -52ms^{-1}.

3. (a) (i) 784 (ii) 768 (iii) 752 (iv) $800 - 16n$

    (b) $a = 800$, $d = -16$. When $u_n = 0$ cars pass each other, $u_n = 800 - 16n = 0$, so $n = 50$ (or, $a = 784$, $d = -16 \Rightarrow 784 - (n-1)16 = 0$). Therefore cars pass each other after 50 seconds.

    (c) As the cars are approaching each other at 16ms^{-1} and need to travel 800m before passing each other, we can use the formula relating distance, constant velocity and time : time = distance/speed. Therefore time = 800/16 = 50 seconds.

4. (a) The nested carts can be modeled by an arithmetic sequence in which $a = 0.95$, $d = 1.13 - 0.95 = 0.18$ and the nth term is the length of n nested carts. So the length of 12 nested carts is $u_{12} = 0.95 + 11 \times 0.18 = 2.93\text{m}$.

    (b) Assume that a nested row of carts is stored lengthwise in the storage space. Then, since the width of a cart is 0.48m and the width of the storage space is 3m, there is space for 6 rows of nested carts. The maximum length of each row is 5m so we require n such that $u_n \leq 5$
    $\Rightarrow 0.95 + (n-1)0.18 \leq 5 \Rightarrow 0.18(n-1) \leq 4.05 \Rightarrow n \leq 23.5$ but $n \in \mathbb{Z}^+$ so $n = 23$. Therefore

the total number of carts which can be stored is $6 \times 23 = 138$.

If the carts are stored widthwise in the storage space there is room for 10 rows. The maximum length of each row is 3m so we require n such that $u_n \leq 3$
$\Rightarrow 0.95 + (n-1)0.18 \leq 3 \Rightarrow n \leq 12.4$ but $n \in \mathbb{Z}^+$ so $n = 12$. Therefore the total number of carts which can be stored is $12 \times 10 = 120$.

Therefore it is more efficient to store the carts lengthwise and the total number of carts which can be stored is 138.

5. (a) $a = 20$, $d = -2$

(b) $\$2 = 8 \times \$0.25$ so the time that eight 25¢ coins will buy is the sum of the first 8 terms of the series which is $S_8 = \frac{8}{2}(2 \times 20 + 7 \times -2) = 4(40 - 14) = 108$. 108 minutes is 1 hour, 44 minutes.

(c) $S_{10} = \frac{10}{2}(2 \times 20 + 9 \times -2) = 5(40 - 18) = 110$. So the maximum parking time is 1 hour 50 minutes.

6. (a) $a = 300$, $d = 300 \times 0.04 = 12$. The value of his investment after 5 years is represented by the sixth term of the arithmetic sequence and $u_6 = a + 5d = 360$. Therefore the value of his investment after 5 years is $360.

(b) The bonds purchased in January 2000 have value $360. The bonds purchased in January 2001 have value $348 and so on. Therefore the value of each years' bonds forms an arithmetic sequence whose first term is 360 and whose common difference is -12. Thus, the total value of the investment after 5 years is, in dollars, $S_5 = \frac{5}{2}(2 \times 360 + (5-1)(-12)) = 1680$.

7. (a) $u_1 = 10$, $d = 1$ and $u_{50} = a + 49d = 10 + 49 = 59 \Rightarrow$ length of longest rod is 59cm.

(b) $\sum_{r=31}^{50} u_r = u_{31} + u_{32} + \ldots + u_{50} = 40 + 41 + \ldots + 59$. This is an arithmetic series with $a = 40$, $d = 1$. Therefore, $S_{20} = 10(2 \times 40 + 19 \times 1) = 990$ and the maximum length of 20 rods is 990cm.

(c) $u_{15} = 43 = a + 14 \times 1 \Rightarrow a = 29$. Therefore $S_{15} = \frac{15}{2}(2 \times 29 + 14 \times 1) = 540$ and so the length of the rods is 540cm.

(d) For a minimum number of rods we need to use the longest rods, so, as $u_n = 59$ and $d = 1$, we have $a + (n-1)1 = 59 \Rightarrow a = 60 - n$. $S_n = \frac{n}{2}(2a + (n-1)1) = 495$ so
$S_n = \frac{n}{2}(2(60-n) + n - 1) = \frac{n}{2}(119 - n) = 495 \Rightarrow n^2 - 119n + 990 = 0$

$\Rightarrow (n-9)(n-110)=0 \Rightarrow n=9, 110$ but $n<50$ so $n=9$ and the minimum number of rods is 9.

8. (a) $a=0, d=8$ (b) $u_r = 8(r-1)$ seconds.

(c) Times measured in seconds.
Machine 1: 60 (to load) + 8 (to go up) + 60 (to unload) +8 (to go down) = 136
Machine 2: $60+8\times 2+60+8\times 2 = 152$
Machine 3: $60+8\times 3+60+8\times 3 = 168$
These times form an arithmetic sequence in which $a=136, d=16$. Since there are 24 machines, one for each floor 2nd and above, $n=24$.
$S_{24} = 12(2\times 136 + 23\times 16) = 7680$ seconds = 128 minutes = 2 hours 8 minutes.

9. (a) $a=1000, d=10$ (b) $u_6 = a+5d = 1000+600 = 1600$

(c) (i) $\left(A+\dfrac{A}{100}\right)+A = 2A+\dfrac{A}{100}$ (ii) $\left(A+\dfrac{2A}{100}\right)+\left(A+\dfrac{A}{100}\right)+A = 3A+\dfrac{3A}{100}$

(d) $\left(A+\dfrac{4A}{100}\right)+\left(A+\dfrac{3A}{100}\right)+\left(A+\dfrac{2A}{100}\right)+\left(A+\dfrac{A}{100}\right)+A = 5A+\dfrac{10A}{100}$

(e) $12A+\left(\dfrac{(1+2+3+4+\ldots+11)A}{100}\right) = 12A+\dfrac{66A}{100}$

(f) $nA+\dfrac{1}{2}n(n-1)\dfrac{A}{100} = nA+\dfrac{n(n-1)A}{200}$

Exercise 1.4

1. (a) 1, 3, 9, 27, 81 (b) 32, 8, 2, $\dfrac{1}{2}$, $\dfrac{1}{8}$ (c) 3, −2, $\dfrac{4}{3}$, $-\dfrac{8}{9}$, $\dfrac{16}{27}$ (d) $\dfrac{4}{9}$, $\dfrac{2}{3}$, 1, $\dfrac{3}{2}$, $\dfrac{9}{4}$

2. (a) $\dfrac{1}{125}, \dfrac{1}{625}, \dfrac{1}{3125}$; $u_n = \left(\dfrac{1}{5}\right)^{n-1}$ (b) 54, 162, 486; $u_n = 2\times 3^{n-1}$

(c) $\dfrac{144}{125}, -\dfrac{864}{625}, \dfrac{5184}{3125}$; $u_n = -\dfrac{2}{3}\left(-\dfrac{6}{5}\right)^{n-1}$ (d) 1.331, 1.4641, 1.61051; $u_n = 1.1^{n-1}$

(e) $8k, -16k, 32k$; $u_n = (-k)(-2)^{n-1}$ (f) $-\dfrac{1}{8}k^4, \dfrac{1}{16}k^5, -\dfrac{1}{32}k^6$; $u_n = k\left(-\dfrac{1}{2}k\right)^{n-1}$

3. (a) $1 \to X, 2X \to X$; $n = 18$ (b) $3 \to X, \dfrac{11}{3}X \to X$; $n = 11$

   (c) $1 \to X, 1.06X \to X$; $n = 29$ (d) $\dfrac{1}{2} \to X, \dfrac{4}{3}X \to X$; $n = 12$

   (e) $1 \to X, 1.08X \to X$; $n = 40$

4. (a) $a = 1, r = 2$, $S_{10} = \dfrac{1(2^{10} - 1)}{2 - 1} = 1023$ (b) $a = 3, r = \dfrac{11}{3}$, $S_5 = \dfrac{20101}{27}$

   (c) $a = 1, r = 1.06$, $S_{25} = 54.864512 = 54.86$, correct to 2 decimal places

   (d) $a = \dfrac{1}{2}, r = \dfrac{4}{3}$, $S_6 = \dfrac{3367}{486}$

   (e) $a = 1, r = \dfrac{1}{2}$, $S_{20} \approx 1.999998093 = 2.00$, correct to 2 decimal places

5. (a) $a = 18, r = \dfrac{2}{3}$ so $S_\infty = \dfrac{18}{1 - 2/3} = 54$ (b) $\dfrac{3}{4}$ (c) $\dfrac{25}{3}$

6. (a) $1.01010101\ldots = 1 + 0.01 + 0.0001 + \ldots = 1 + \dfrac{1}{100} + \dfrac{1}{10000} + \ldots$ Therefore, $a = 1$,

   $r = \dfrac{1}{100}$ and $S_\infty = 1.01010101\ldots = \dfrac{1}{99/100} = \dfrac{100}{99}$.

   (b) (i) $0.44444\ldots = \dfrac{4}{10} + \dfrac{4}{100} + \dfrac{4}{1000} + \ldots$ so $a = \dfrac{4}{10}, r = \dfrac{1}{10}, S_\infty = \dfrac{4/10}{1 - 1/10} = \dfrac{4}{9}$.

      (ii) $0.91919191\ldots = \dfrac{91}{100} + \dfrac{91}{10000} + \ldots$ so $a = \dfrac{91}{100}, r = \dfrac{1}{100}, S_\infty = \dfrac{91}{99}$.

      (iii) $47.3473473\ldots = 47.3 + 0.0473 + 0.0000473 + \ldots = 47.3\left(1 + \dfrac{1}{1000} + \dfrac{1}{1000000} + \ldots\right)$

      Therefore $a = 1, r = \dfrac{1}{1000}, S_\infty = 47.3\left(\dfrac{1}{1 - 1/1000}\right) = \dfrac{47300}{999}$.

7. (a) $u_5 = 64, u_8 = 8 \Rightarrow ar^4 = 64, ar^7 = 8 \Rightarrow r^3 = \tfrac{1}{8} \Rightarrow r = \tfrac{1}{2}$. So first term is $a = 64 \times 16 = 1024$ and common ratio is $\dfrac{1}{2}$.

   (b) $a = 20, u_{37} = ar^{36} = 30 \Rightarrow r^{36} = 1.5 \Rightarrow r = 1.0113266 = 1.0113$, correct to 4 decimal places

8. $S_n = \frac{1}{2}(3^n - 1)$. Therefore, $S_1 = a = \frac{1}{2}(3^1 - 1) = 1$, $S_2 = a + ar = \frac{1}{2} \times 8 = 4 \Rightarrow 1 + r = 4 \Rightarrow r = 3$ and $u_5 = 1 \times 3^4 = 81$.

9. (a) $a = 13.5$, $r = \frac{1}{3}$, $u_n = \frac{1}{162} \Rightarrow 13.5\left(\frac{1}{3}\right)^{n-1} = \frac{1}{162} \Rightarrow \left(\frac{1}{3}\right)^{n-1} = \frac{1}{13.5 \times 162} = \frac{1}{2187}$. Therefore, $3^{n-1} = 2187 \Rightarrow 3^{n-1} = 3^7 \Rightarrow n = 8$.

 (b) $a = 10$, $r = -2$, $u_n = ar^{n-1} = -5120 \Rightarrow 10(-2)^{n-1} = -5120 \Rightarrow (-2)^{n-1} = -512$
 $\Rightarrow (-2)^{n-1} = -2^9 \Rightarrow n = 10$.

10. (a) $S_6 = \frac{a(r^6 - 1)}{r - 1} = 3$, $S_{10} = \frac{a(r^{10} - 1)}{r - 1} = 18 \Rightarrow a(r^6 - 1) = 3r - 3$ .... (1) and
 $a(r^{10} - 1) = 18r - 18$ ...... (2)

 Equations (1) and (2) can be solved simultaneously by division:

 $$\frac{a(r^{10} - 1)}{a(r^6 - 1)} = \frac{18(r - 1)}{3(r - 1)} \Rightarrow r^{10} - 1 = 6(r^6 - 1) \Rightarrow r^{10} - 6r^6 + 5 = 0$$

 (b) Use the graphing calculator to solve $r^{10} - 6r^6 + 5 = 0$. Therefore, $r = 1$, $r = -1$, $r = 1.5400512$, $r = -1.5400512$. The trivial solutions are $r = 1$ and $r = -1$ because they give rise to geometric series whose terms are all equal ($r = 1$) or whose terms alternate with equal magnitude ($r = -1$). So, the non-trivial solutions are $r = -1.54, 1.54$.

Exercise 1.5

1. Working in millions, the situation can be modeled by a geometric sequence with $a = 280$, $r = 1 + \frac{1.1}{100} = 1.011$. Therefore,

 (i) in 2 years time, $u_3 = 280(1.011^2) = 286.19388$, so the population will be about 286 million.

 (ii) in 10 years time, population will be $u_{11} = 280(1.011)^{10} \approx 312$ million.

 (iii) in 20 years time, population will be $u_{21} = 280(1.011)^{20} \approx 348$ million.

2. Vladimir's investment can be modeled by a geometric sequence with $a = 50\,000$ and $r = 1 + \dfrac{5.6}{100} = 1.056$. After 5 years, $u_6 = 50000(1.056)^5 = 65658.29$ and so Vladimir's investment is worth $65658.29. Galina's investment can be modeled by a geometric sequence with $a = 15000$ and $r = 1 + \dfrac{4.7}{100} = 1.047$. After 20 years, $u_{21} = 15000(1.047)^{20} = 37585.89$, and so her investment is worth $37585.89.

3. (a) $14000(1.045) = 14630$, so 630 antelope need to be culled each year.

   (b) $34000(1.04)^6 = 43020.84$, so that 6 years later the population is about 43 000.

4. (i) $P(1.07)^n = 2P \Rightarrow 1.07^n = 2$. Using the method of Example 1.1: $1 \to X$, $1.07X \to X$ which gives $X = 1.96715$ when $n = 10$ and $X = 2.10485$ when $n = 11$, so that it will take almost 11 years for the population to double.

   (ii) The time period between the beginning of 2001 and the beginning of 2006 is 5 years. Therefore the value of the stamp is given by $u_6 = 65000(1.12)^5 \approx 11455.22$ which is about $11 500.

   (iii) In January 2004, 10 units $\equiv$ $1. One year later, $10 \equiv \$1 \times 0.65$ so that 6 years later 10 units $= \$1 \times 0.65^6 = \$0.07541889$. Therefore, at this time $1 is worth 132.59 units of the currency.

5. This situation is modeled by a geometric sequence with $a = 38000$ and $r = 1 - \dfrac{23}{100} = 0.77$. Peter sells the vehicle after 5 years, so we require $u_6 = 38000(0.77)^5 = 10285.78$. Therefore the Gladiator's value is $10300.

6. (a) (i) $\dfrac{4.8}{12} = 0.4\%$ per month.   (ii) $\dfrac{4.8}{365} \approx 0.0131507\%$ per day.

   (b) (i) The situation is modeled by a geometric sequence with $a = 1000$, $r = 1 + \dfrac{.4}{100} = 1.004$. Five years is 60 months, and therefore you require $u_{61} = 1000(1.004)^{60} = 1270.64$ and the value of the investment is $1270.64.

   (ii) The geometric sequence has $a = 1000$, $r = 1 + \dfrac{0.0131507}{100} = 1.000131507$, and 5 years is 1825 days (neglecting Leap Years) so you require $u_{1826} = 1000(1.000131507)^{1825} = $ 1271.23. Note that it is important to give the common ratio, r, very accurately, in order to be able to give a final answer correct to the nearest cent. If you had used

$r = 1.000132$ you would have obtained a final answer of $1272.37. $1.14 over 5 years might not seem very much, but financial institutions need to work to the nearest cent.

7. This can be modeled by a geometric <u>series</u> with $a = 18$, working in billions, and $r = 1 + \frac{8}{100} = 1.08$. You need to find n so that $S_n = 1000$. $S_n = \frac{18(1.08^n - 1)}{1.08 - 1} = 1000$
$\Rightarrow 1.08^n - 1 = 4.4444 \Rightarrow 1.08^n = 5.4444$. Using the method of Example 1.1, $1 \to X$, $1.08X \to X$ gives $1.08^{22} = 5.4365$ and $1.08^{23} = 5.8715$ so that n is very close to 22 and the mineral resource will become exhausted after 22 years.

8. (a) The situation is modeled by a geometric sequence with $a = 1000$, $r = 1 - \frac{10}{100} = 0.9$.
The value of the first machine, in dollars, at the time of bankruptcy is given by
$u_6 = 1000(0.9)^5 = 590.49$.

(b) The machine bought at the start of the second year will have value at bankruptcy given by u_5. The value of the machine bought at the start of the third year will have value at bankruptcy given by u_4 and so on. Therefore, the total value of all 5 machines at bankruptcy will be $u_5 + u_4 + u_3 + u_2 + u_1 = S_5$. Now the first term is $u_1 = 1000 \times 0.9 = 900$ so $a = 900$ and $r = 0.9$.

(c) $S_5 = \frac{900(1 - 0.9^5)}{1 - 0.9} = 3685.59$. Therefore, the value of all five machines, to the nearest $100, at the time of bankruptcy is $3700.

9. The population of rats is given by the terms of a geometric sequence with $a = 160$ (working in thousands) and $r = 1 + \frac{1.7}{100} = 1.017$. The population of mice is given by the terms of a geometric sequence with $a = 70$ and $r = 1 + \frac{3.8}{100} = 1.038$. You need to find n such that
$160(1.017)^n = 70(1.038)^n \Rightarrow \left(\frac{1.038}{1.017}\right)^n = \frac{16}{7} \Rightarrow 1.02065^n = 2.2857 \Rightarrow n = 40.4$.
Using the method of Example 1.1, $1 \to X$, $1.02065X \to X$ gives $1.02065^{40} = 2.265$ and $1.02065^{41} = 2.312$. Therefore, it will take about 40 years for the mice population to exceed the rat population.

10. (a) Interest is 1% per month. Therefore, after 6 months, she owes $1000(1.01)^6$, but then she repays an amount P and so she owes $1000(1.01)^6 - P$.

(b) At the start of the second six month period, she owes $1000(1.01)^6 - P$. Therefore, by the end of the year she owes $\left(1000(1.01)^6 - P\right)(1.01)^6$, at which time she makes the second payment, P. Therefore, as the loan is paid off at the end of the year:

$\left(1000(1.01)^6 - P\right)(1.01)^6 - P = 0 \Rightarrow 1000(1.01)^{12} = P\left(1+(1.01)^6\right)$

$\Rightarrow P(1+1.06152) \approx 1000 \times 1.126825 = 1126.825 \Rightarrow P = \dfrac{1126.825}{2.06152} \approx 546.60$.

Exercise 1.6

1. (i) $(ab)^2 = a^2b^2$. Therefore, $(abc^3) \times (ab)^2 = a^1b^1c^3a^2b^2 = a^3b^3c^3 = (abc)^3$

   (ii) $\sqrt{\dfrac{a^5b^3c}{abc^3}} = \sqrt{\dfrac{a^4b^2}{c^2}} = \left(\dfrac{a^4b^2}{c^2}\right)^{\frac{1}{2}} = \dfrac{a^2b}{c}$

2. (i) $2a^3b^2$ (ii) a^3b^2c (iii) $2a^2b^{-1}c$ $\left(\text{or } \dfrac{2a^2c}{b}\right)$ (iv) a^2bc^{-2} $\left(\text{or } \dfrac{a^2b}{c^2}\right)$

   (v) $a^{-5}b^{-2}$ $\left(\text{or } \dfrac{1}{a^5b^2}\right)$ (vi) $a^{-7}b^{-8}$ $\left(\text{or } \dfrac{1}{a^7b^8}\right)$ (vii) $4^{-3}ab$ $\left(\text{or } \dfrac{ab}{64}\right)$

3. (a) 8 (b) 32 (c) $\dfrac{4}{9}$ (d) 2 (e) 9 (f) $\dfrac{2}{5}$

   (g) $\dfrac{625}{256}$ (h) 16 (i) $\dfrac{8}{27}$ (j) $\dfrac{27}{100}$ (k) $\dfrac{24}{5}$

4. (a) 2 (b) 1 (c) $\dfrac{3}{7}$ (d) $-\dfrac{1}{6}$ (e) $\dfrac{13}{8}$ (f) $-\dfrac{1}{2}$ (g) $\dfrac{2}{3}$ (h) 2

Exercise 1.7

1. (a) $x = \log_2 3$ (b) $x = \log_3 2$ (c) $x = \log_6 5$ (d) $x = \log_{10} 7 - 1$ (e) $x = \log_5 4$

   (f) $x = \dfrac{1}{2}\log_3\left(\dfrac{7}{4}\right)$ (g) $x = \log_2 9 + 3$ (h) $x = \log_3\left(\dfrac{5}{2}\right)$ (i) $x = \dfrac{1}{\log_3 11}$

   (j) $x = \dfrac{1}{3}(2 - \log_4 12)$ (k) $x = \pm\sqrt{\log_6 3}$

2. (a) 9 (b) 64 (c) $x = 9^{\frac{3}{2}} = \left(9^{\frac{1}{2}}\right)^3 = 27$ (d) $2x = 6^3 = 216 \Rightarrow x = 108$

   (e) $x+1 = 32 \Rightarrow x = 31$ (f) $\frac{1}{x} = 5^3 \Rightarrow x = \frac{1}{125}$ (g) $\frac{x-1}{2} = 2^7 = 128 \Rightarrow x-1 = 256 \Rightarrow x = 257$

   (h) $\sqrt{x} = 4^2 = 16 \Rightarrow x = 256$

3. (a) $2u + 3v$ (b) $u - v$ (c) $4u - v$ (d) $2u - v$

4. (a) $x = 2.69$ (b) $x = 1.26$ (c) $x = 0.112$ (d) $x = \log_{10} 3.1 + 6 = 6.49$

   (e) $x - 1 = \log_{10} 2097 = 3.32160 \Rightarrow x = 4.32$ (f) $10^x = 4 - 2.196 = 1.084 \Rightarrow x = 0.0350$

   (g) $100^x = 10^{2x} = 78.5 \Rightarrow 2x = \log_{10} 78.5 \Rightarrow x = \frac{1}{2}\log_{10} 78.5 = 0.947$

   (h) $0.1^x = 10^{-x} = 19 \Rightarrow -x = \log_{10} 19 \Rightarrow x = -1.28$

5. (a) $\log 6$ (b) $\log 5$ (c) $\log 6$ (d) $\log \frac{2}{3}$ (e) $\log 3$

   (f) $2\log 5$ (or $\log 25$) (g) $\log 6$ (h) $\log 72$ (i) $\log 2$

   (j) $\log \frac{3}{8}$ (k) $\log 8a^2$

6. (a) 1 (b) 4 (c) $\frac{1}{2}$ (d) 0 (e) -3 (f) -1 (g) 4 (h) -8 (i) $\frac{5}{2}$

7. $\log_2(x^3 y^2) = 28 \Rightarrow 3\log_2 x + 2\log_2 y = 28$, $\log_2\left(\frac{x^4}{y^3}\right) = -8 \Rightarrow 4\log_2 x - 3\log_2 y = -8$
   $\Rightarrow 3u + 2v = 28$, $4u - 3v = -8 \Rightarrow u = 4$, $v = 8 \Rightarrow \log_2 x = 4$, $\log_2 y = 8$. Therefore,
   $x = 2^4 = 16$, $y = 2^8 = 256$.

Exercise 1.8

1. (a) $2^x = 5 \Rightarrow x = \log_2 5 = \frac{\log_{10} 5}{\log_{10} 2} = 2.32$ (b) $3^x = 88 \Rightarrow x = \log_3 88 = \frac{\log_{10} 88}{\log_{10} 3} = 4.08$

(c) $4^{x+1} = 15 \Rightarrow x+1 = \log_4 15 = \dfrac{\log_{10} 15}{\log_{10} 4} = 1.95345 \Rightarrow x = 0.953$

(d) $2^{3-x} = 22 \Rightarrow 3-x = \log_2 22 \Rightarrow x = 3 - \dfrac{\log_{10} 22}{\log_{10} 2} = -1.4594 \Rightarrow x = -1.46$

(e) $5^{x-2} = 338 \Rightarrow x-2 = \log_5 338 = \dfrac{\log_{10} 338}{\log_{10} 5} = 3.61806 \Rightarrow x = 5.62$

(f) $3^{-x^2} = \dfrac{1}{2} \Rightarrow -x^2 = \log_3 \dfrac{1}{2} = \dfrac{\log_{10} \frac{1}{2}}{\log_{10} 3} = -0.6309298 \Rightarrow x = \pm 0.794$

2. (a) $\log_4 25 = \dfrac{\log_2 25}{\log_2 4} = \dfrac{\log_2 25}{2 \log_2 2} = \dfrac{1}{2} \log_2 25$ so $\log_2 x = \dfrac{\log_2 25}{2} = \log_2 5 \Rightarrow x = 5$

(b) $\log_3 x = \dfrac{\log_3 (7x-6)}{\log_3 9} = \dfrac{1}{2} \log_3 (7x-6) \Rightarrow x = \sqrt{7x-6} \Rightarrow x^2 = 7x-6 \Rightarrow x^2 - 7x + 6 = 0$

$\Rightarrow (x-6)(x-1) = 0 \Rightarrow x = 1, 6$

(c) $\log_8 32x = \dfrac{\log_2 32x}{\log_2 8} = \dfrac{1}{3} \log_2 32x$ so $\log_2 x + \dfrac{1}{3} \log_2 32x = 1 \Rightarrow 3 \log_2 x + \log_2 32x = 3$

$\Rightarrow \log_2 32x^4 = 3 \Rightarrow 32x^4 = 2^3 = 8 \Rightarrow x^4 = \dfrac{1}{4} \Rightarrow x = \pm \dfrac{1}{\sqrt{2}}$, but $x > 0$. Therefore, $x = \dfrac{1}{\sqrt{2}}$.

(d) $\log_x 81 = \log_3 x \Rightarrow \dfrac{\log_3 81}{\log_3 x} = \log_3 x \Rightarrow 4 = (\log_3 x)^2 \Rightarrow \log_3 x = \pm 2 \Rightarrow x = 3^{\pm 2} \Rightarrow x = 3^2 = 9$

and $x = 3^{-2} = \dfrac{1}{9}$.

(e) $\log_2 (2x+1) - \dfrac{\log_2 x}{\log_2 4} = \log_2 3 \Rightarrow \log_2 (2x+1) - \dfrac{1}{2} \log_2 x = \log_2 3 \Rightarrow \dfrac{2x+1}{\sqrt{x}} = 3 \Rightarrow 2x+1$

$= 3\sqrt{x} \Rightarrow (2x+1)^2 = 9x \Rightarrow 4x^2 + 4x + 1 = 9x \Rightarrow 4x^2 - 5x + 1 = 0$

$\Rightarrow (4x-1)(x-1) = 0 \Rightarrow x = \dfrac{1}{4}, 1$

(f) $\dfrac{\log_2 x^2}{3} = \dfrac{\log_2 8x}{2} \Rightarrow 2 \log_2 x^2 = 3 \log_2 8x \Rightarrow \log_2 x^4 = \log_2 512 x^3 \Rightarrow x^4 - 512 x^3 = 0$

$\Rightarrow x^3 (x - 512) = 0$ and as $x > 0$, $x = 512$.

3. (a) $\dfrac{\log_{10} x}{\log_{10} 2} = \dfrac{\log_{10} 3}{\log_{10} 5} \Rightarrow \log_{10} x = 0.2054849 \Rightarrow x = 1.61$

(b) $\dfrac{\log_{10} x}{\log_{10} 7} = \dfrac{\log_{10} 9}{\log_{10} 2} \Rightarrow \log_{10} x = 2.678897 \Rightarrow x = 477$

(c) $\dfrac{\log_{10} x}{\log_{10} 2} + \dfrac{\log_{10} x}{\log_{10} 4} = \dfrac{\log_{10} 5}{\log_{10} 2} \Rightarrow \dfrac{\log_{10} x}{\log_{10} 2} + \dfrac{\log_{10} x}{2\log_{10} 2} = \dfrac{\log_{10} 5}{\log_{10} 2} \Rightarrow \dfrac{3}{2}\log_{10} x = \log_{10} 5$

$\Rightarrow x^{\frac{3}{2}} = 5 \Rightarrow x = 5^{\frac{2}{3}} = 2.92$

(d) $\dfrac{\log_{10} x}{\log_{10} 3} = \log_{10} 32 \Rightarrow \log_{10} x = 0.718139 \Rightarrow x = 5.23$

4. (a) $x > \dfrac{\log_{10} 5000}{\log_{10} 2} = 12.3$ (b) $x > \dfrac{\log_{10} 697}{\log_{10} 3} = 5.96$ (c) $x > \dfrac{1}{\log_{10} 1.01} = 231$

(d) $x+1 > \dfrac{\log_{10} 33}{\log_{10} 1.15} = 25.01759 \Rightarrow x > 24.0$

(e) $x\log_{10} 0.99 < \log_{10} 0.5 \Rightarrow x > \dfrac{\log_{10} 0.5}{\log_{10} 0.99} = 69.0$. (Note that $\log_{10} 0.99 < 0$ so that the inequality sign must change when the inequality is divided by $\log_{10} 0.99$.)

(f) $x\log_{10} 0.5 < \log_{10} 0.001 \Rightarrow x > \dfrac{\log_{10} 0.001}{\log_{10} 0.5} = 9.97$. (Note that $\log_{10} 0.5 < 0$ so that the inequality sign must change when the inequality is divided by $\log_{10} 0.5$.)

Exercise 1.9

1. (a) $(x+y)^4 = x^4 + 4x^3 y + 6x^2 y^2 + 4xy^3 + y^4$ (b) $(1+y)^3 = 1 + 3y + 3y^2 + y^3$

(c) $(1+x)^7 = 1 + 7x + 21x^2 + 35x^3 + 35x^4 + 21x^5 + 7x^6 + x^7$

(d) $(1-x)^6 = 1 - 6x + 15x^2 - 20x^3 + 15x^4 - 6x^5 + x^6$

(e) $(k-1)^8 = k^8 - 8k^7 + 28k^6 - 56k^5 + 70k^4 - 56k^3 + 28k^2 - 8k + 1$

2. (a) $(9x^2)(16x^4) = 144x^6$ (b) $(8a^3)(25b^2) = 200a^3 b^2$ (c) -1152

(d) $\left(\dfrac{x^3}{8}\right)(4y^2) = \dfrac{1}{2}x^3 y^2$ (e) $8x^3$ (f) $\left(\dfrac{1}{81}a^4\right)\left(-\dfrac{27}{8}b^3\right) = -\dfrac{1}{24}a^4 b^3$

3. (a) $x^4 + 4x^3(2y) + 6x^2(2y)^2 + 4x(2y)^3 + (2y)^4 = x^4 + 8x^3 y + 24x^2 y^2 + 32xy^3 + 16y^4$

(b) $1 + 5(2x) + 10(2x)^2 + 10(2x)^3 + 5(2x)^4 + (2x)^5 = 1 + 10x + 40x^2 + 80x^3 + 80x^4 + 32x^5$

(c) $2^3 + 3(2^2)\left(-\dfrac{k}{2}\right) + 3(2)\left(-\dfrac{k}{2}\right)^2 + \left(-\dfrac{k}{2}\right)^3 = 8 - 6k + \dfrac{3k^2}{2} - \dfrac{k^3}{8}$

(d) $3^7 + 7(3^6)(2x) + 21(3^5)(2x)^2 + 35(3^4)(2x)^3 + 35(3^3)(2x)^4 + 21(3^2)(2x)^5 + 7(3)(2x)^6$
$+ (2x)^7 = 2187 + 10206x + 20412x^2 + 22680x^3 + 15120x^4 + 6048x^5 + 1344x^6 + 128x^7$

(e) $1 + 4\left(\dfrac{2}{3}a\right) + 6\left(\dfrac{2}{3}a\right)^2 + 4\left(\dfrac{2}{3}a\right)^3 + \left(\dfrac{2}{3}a\right)^4 = 1 + \dfrac{8}{3}a + \dfrac{8}{3}a^2 + \dfrac{32}{27}a^3 + \dfrac{16}{81}a^4$

(f) $\left(\dfrac{1}{3}x\right)^3 + 3\left(\dfrac{1}{3}x\right)^2\left(\dfrac{3}{2}y\right) + 3\left(\dfrac{1}{3}x\right)\left(\dfrac{3}{2}y\right)^2 + \left(\dfrac{3}{2}y\right)^3 = \dfrac{x^3}{27} + \dfrac{x^2 y}{2} + \dfrac{9xy^2}{4} + \dfrac{27y^3}{8}$

(g) $(x^2)^4 + 4(x^2)^3\left(-\dfrac{1}{x}\right) + 6(x^2)^2\left(-\dfrac{1}{x}\right)^2 + 4(x^2)\left(-\dfrac{1}{x}\right)^3 + \left(-\dfrac{1}{x}\right)^4 = x^8 - 4x^5 + 6x^2 - \dfrac{4}{x} + \dfrac{1}{x^4}$

(h) $\left(\dfrac{x}{2}\right)^6 + 6\left(\dfrac{x}{2}\right)^5\left(\dfrac{2}{x}\right) + 15\left(\dfrac{x}{2}\right)^4\left(\dfrac{2}{x}\right)^2 + 20\left(\dfrac{x}{2}\right)^3\left(\dfrac{2}{x}\right)^3 + 15\left(\dfrac{x}{2}\right)^2\left(\dfrac{2}{x}\right)^4 + 6\left(\dfrac{x}{2}\right)\left(\dfrac{2}{x}\right)^5 + \left(\dfrac{2}{x}\right)^6$
$= \dfrac{x^6}{64} + \dfrac{3x^4}{8} + \dfrac{15x^2}{4} + 20 + \dfrac{60}{x^2} + \dfrac{96}{x^4} + \dfrac{64}{x^6}$

4. (a) 36   (b) 364   (c) 924   (d) 190   (e) 6435

5. (a) $\binom{10}{3} = 120$ (b) $\binom{18}{11} = 31824$ (c) $\binom{14}{9}(-1)^9 = -2002$ (d) $\binom{21}{6} = 54264$

(e) $\binom{15}{10}(2^5) = 3003 \times 32 = 96096$ (f) $\binom{9}{5}2^5 = 126 \times 32 = 4032$

(g) $\binom{16}{6}2^{10}\left(-\dfrac{1}{2}\right)^6 = 8008 \times 2^4 = 128128$ (h) $\binom{11}{4}3^7 \times 2^4 = 330 \times 2187 \times 16 = 11547360$

6. (a) $\binom{9}{3}(3x)^3 = 84 \times 27x^3 = 2268x^3$ (b) $\binom{13}{1}2^{12}(-x) = -13 \times 4096x = -53248x$

(c) $\binom{25}{2}x^2 = 300x^2$ (d) $\binom{19}{16}(-x)^{16} = 969x^{16}$ (e) $\binom{5}{2}a^3\left(\frac{b}{3}\right)^2 = \frac{10a^3b^2}{9}$

(f) $\binom{11}{4}3^7\left(\frac{1}{2}x\right)^4 = \frac{330 \times 2187}{16}x^4 = \frac{360855}{8}x^4$ (g) $\binom{7}{2}(3x)^5\left(-\frac{1}{3}\right)^2 = 567x^5$

(h) $\binom{6}{4}(2x)^2\left(\frac{y}{2}\right)^4 = \frac{15}{4}x^2y^4$

7. (a) $(1+x)^4 = 1 + 4x + 6x^2 + 4x^3 + x^4 = 1 + 0.4 + 0.06 + 0.004 + 0.0001 = 1.4641$

(b) $(1+x)^7 = 1 + 7x + 21x^2 + 35x^3 + 35x^4 + 21x^5 + 7x^6 + x^7$
$= 1 + 7 \times 10^{-2} + 21 \times 10^{-4} + 35 \times 10^{-6} + 35 \times 10^{-8} + 21 \times 10^{-10} + 7 \times 10^{-12} + 10^{-14}$
$= 1.07213535210701$

(c) $1.01^8 = 1.0828567056280801$, the digits being the appropriate row of Pascal's triangle.

8. (a) $\binom{5}{4}\frac{1}{k}(kx)^4 = 625x^4 \Rightarrow k^3 = 125 \Rightarrow k = 5$

(b) $\binom{7}{3}k^4(-x)^3 = -560x^3 \Rightarrow -35k^4 = -560 \Rightarrow k^4 = 16 \Rightarrow k = \pm 2$ but $k > 0$. Therefore, $k = 2$.

(c) $\binom{12}{7}(kx)^7 = \frac{11264}{243}x^7 \Rightarrow 792k^7 = \frac{11264}{243} \Rightarrow k^7 = \frac{128}{2187} \Rightarrow k = \frac{2}{3}$

9. (a) $(1+y)^4 = 1 + 4y + 6y^2 + 4y^3 + y^4$

(b) $y = x + x^2 \Rightarrow y^2 = (x+x^2)^2 = x^2 + 2x^3 + x^4$, $y^3 = (x+x^2)^3 = x^3 + 3x^4 + 3x^5 + x^6$
$y^4 = (x+x^2)^4 = x^4 + 4x^5 + 6x^6 + 4x^7 + x^8$. Therefore, $(1+x+x^2)^4 = 1 + 4(x+x^2) +$
$6(x+x^2)^2 + 4(x+x^2)^3 + (x+x^2)^4 = 1 + 4x + 10x^2 + 16x^3 + 19x^4 + 16x^5 + 10x^6 + 4x^7 + x^8$.

Solutions to Unit 2 Exercises

Exercise 2.1

1. (a) $x = 1$ has no image because $\frac{2}{0}$ is undefined.

   (b) x does not have a unique image because each value of x is assigned to two values in the range.

   (c) x has no image for all values of $x > \frac{2}{3}$, $x \in \mathbb{R}$.

   (d) $n = 1$ and $n = 2$ have no image.

   (e) n is mapped to more than one image. For example, $10 \to 1, 2, 5, 10$.

   (f) $x = 0.5$ is mapped to 0 and 1.

2. (a) 1   (b) -2 (c) -14

3. (a) 9   (b) 0   (c) $-\frac{336}{25}$ (d) $4\sqrt{10} - 2 \approx 10.6$

4. (a) $-\frac{2}{9}$ (b) $\frac{3}{4}$ (c) $\frac{11}{100}$

5. (a) $-7 \leq f(x) \leq 13$ (b) $-5 < f(x) < 5$ (c) $0 \leq f(x) \leq 100$ (d) $-1 < f(x) \leq 1$

6. (a) 0.00366   (b) 1.73   (c) 0.699   (d) 0.803

7. (a) 7   (b) 3   (c) 0   (d) 6   (e) -3

Exercise 2.2

1. (a) 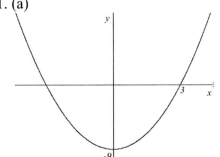   (b) 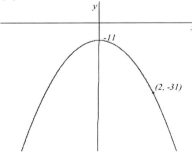   (c)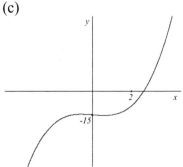

(d) 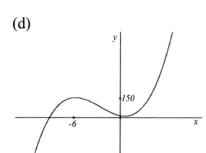 (e) 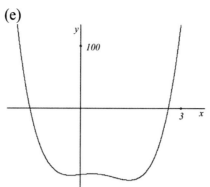 (f)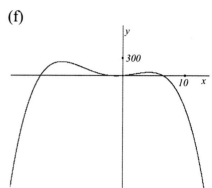

2. (a) 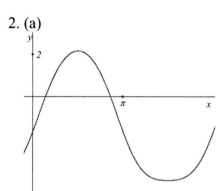 (b) 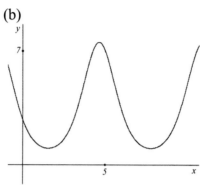 (c)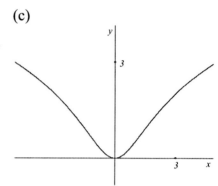

(d) 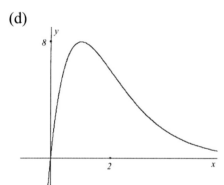 (e) 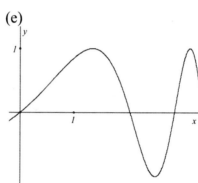 (f)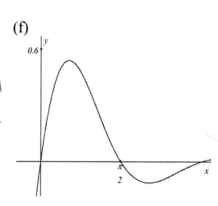

(g) 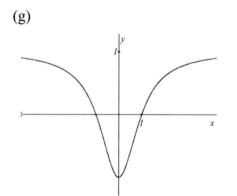 (h)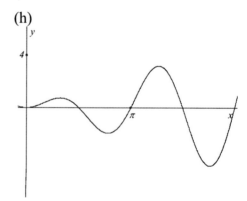

3. (a)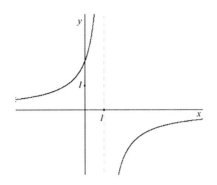

(i) $x = 1$ (ii) $y = 0$

(b)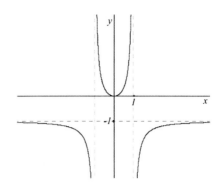

(i) $x = \pm 1$ (ii) $y = -1$

(c)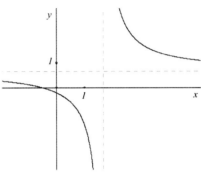

(i) $x = \dfrac{5}{3}$ (ii) $y = \dfrac{2}{3}$

(d)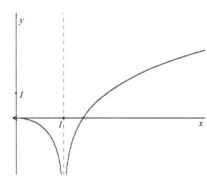

(i) $x = 1$ (ii) None

(e)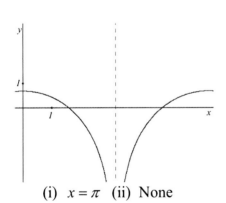

(i) $x = \pi$ (ii) None

(f)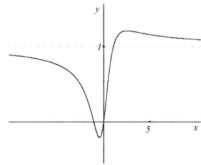

(i) None (ii) $y = 1$

Exercise 2.3

1. A translation of $\begin{pmatrix} 2 \\ 4 \end{pmatrix}$ (translation of 2 units in the positive x direction and 4 units in the positive y direction).

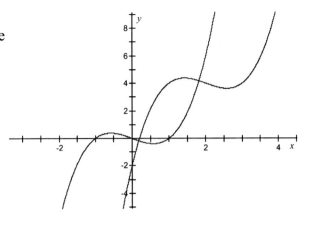

2. (a) A reflection in $y = 0$ (x-axis).

   (b) A translation of $\begin{pmatrix} 2 \\ 3 \end{pmatrix}$ (translation of 2 units in the positive x direction and 3 units in the positive y direction).

   (c) A stretch of factor 2 parallel to the x-axis or stretch of factor $\dfrac{1}{4}$ parallel to the y-axis.

   (d) A translation of $\begin{pmatrix} -4 \\ 0 \end{pmatrix}$ (translation of 4 units in the negative x direction).

3. (a) A translation of $\begin{pmatrix} -3 \\ 1 \end{pmatrix}$.

   (b) A stretch of factor $\dfrac{1}{3}$ parallel to the y-axis or a stretch of factor $\sqrt{3}$ parallel to the x-axis.

   (c) A translation of $\begin{pmatrix} \frac{\pi}{6} \\ 1 \end{pmatrix}$.

   (d) A translation of $\begin{pmatrix} 4 \\ 5 \end{pmatrix}$.

   (e) A reflection in the x-axis followed by a translation of $\begin{pmatrix} 0 \\ 4 \end{pmatrix}$.

   (f) A translation of $\begin{pmatrix} -1 \\ -5 \end{pmatrix}$.

4. (a) A translation of $\begin{pmatrix} 0 \\ -1 \end{pmatrix}$.

(b) A stretch of factor $\frac{1}{3}$ parallel to the x-axis and a translation of $\begin{pmatrix} 0 \\ 1 \end{pmatrix}$.

(c) A translation of $\begin{pmatrix} -4 \\ -4 \end{pmatrix}$.

5. (a)

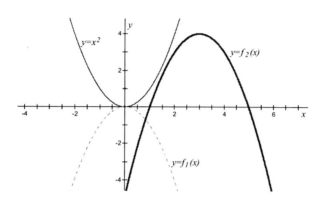

$y = x^2 \Rightarrow y = -x^2 \Rightarrow y = -(x-3)^2 + 4 \Rightarrow y = 4 - (x-3)^2$

(b)

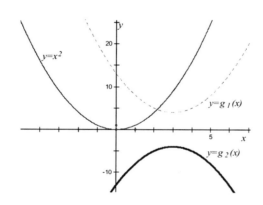

$y = x^2 \Rightarrow y = (x-3)^2 + 4 \Rightarrow y = -\left((x-3)^2 + 4\right) \Rightarrow -4 - (x-3)^2$

Exercise 2.4

1. (a) $(x^2 + 4x + 4) - 4 - 6 = (x+2)^2 - 10$ (b) $(x^2 + 12x + 36) - 36 + 15 = (x+6)^2 - 21$

(c) $(x^2 - 2x + 1) - 1 + 9 = (x-1)^2 + 8$ (d) $(x^2 + 6x + 9) - 9 - 11 = (x+3)^2 - 20$

(e) $(x^2 - 8x + 16) - 16 - 1 = (x-4)^2 - 17$ (f) $(x^2 - 4x + 4) - 4 + 19 = (x-2)^2 + 15$

(g) $(x^2 + 30x + 225) - 225 + 205 = (x+15)^2 - 20$

(a) Check   (b) Check

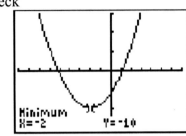 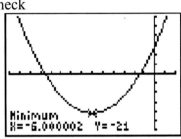

So $x^2 + 4x - 6 = (x-(-2))^2 - 10$ So $x^2 + 12x + 15 = (x-(-6))^2 - 21$
$= (x+2)^2 - 10$ $= (x+6)^2 - 21$

Checks of further parts of this question are omitted.

2. (a) $2 - (x^2 - 4x) = 2 - (x^2 - 4x + 4) + 4 = 6 - (x-2)^2$

(b) $9 - (x^2 - 6x) = 9 - (x^2 - 6x + 9) + 9 = 18 - (x-3)^2$

(c) $6 - (x^2 + 18x) = 6 - (x^2 + 18x + 81) + 81 = 87 - (x+9)^2$

(d) $5 - (x^2 - 16x) = 5 - (x^2 - 16x + 64) + 64 = 69 - (x-8)^2$

(e) $-7 - (x^2 - 4x) = -7 - (x^2 - 4x + 4) + 4 = -3 - (x-2)^2$

(f) $17 - (x^2 + 8x) = 17 - (x^2 + 8x + 16) + 16 = 33 - (x+4)^2$

(g) $-10 - (x^2 + 2x) = -10 - (x^2 + 2x + 1) + 1 = -9 - (x+1)^2$

(a) Check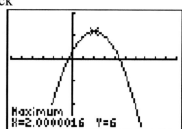

So $2+4x-x^2 = 6-(x-2)^2$

(b) Check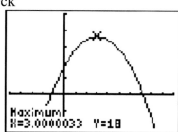

So $9+6x-x^2 = 18-(x-3)^2$

Checks of further parts of this question are omitted.

3. (a) $\left(x^2+3x+\dfrac{9}{4}\right)-\dfrac{9}{4}-5 = \left(x+\dfrac{3}{2}\right)^2 - \dfrac{29}{4}$ (b) $\left(x^2+5x+\dfrac{25}{4}\right)-\dfrac{25}{4}+2 = \left(x+\dfrac{5}{2}\right)^2 - \dfrac{17}{4}$

(c) $\left(x^2-x+\dfrac{1}{4}\right)-\dfrac{1}{4}-11 = \left(x-\dfrac{1}{2}\right)^2 - \dfrac{45}{4}$ (d) $\left(x^2-7x+\dfrac{49}{4}\right)-\dfrac{49}{4}+17 = \left(x-\dfrac{7}{2}\right)^2 + \dfrac{19}{4}$

(e) $\left(x^2+11x+\dfrac{121}{4}\right)-\dfrac{121}{4}-5 = \left(x+\dfrac{11}{2}\right)^2 - \dfrac{141}{4}$

(f) $\left(x^2-15x+\dfrac{225}{4}\right)-\dfrac{225}{4}+32 = \left(x-\dfrac{15}{2}\right)^2 - \dfrac{97}{4}$

(g) $\left(x^2+17x+\dfrac{289}{4}\right)-\dfrac{289}{4}+73 = \left(x+\dfrac{17}{2}\right)^2 + \dfrac{3}{4}$

Checks may be carried out as for questions 1 & 2 and are omitted here.

4. (a) $7-(x^2-3x) = 7-\left(x^2-3x+\dfrac{9}{4}\right)+\dfrac{9}{4} = \dfrac{37}{4} - \left(x-\dfrac{3}{2}\right)^2$

(b) $1-(x^2+x) = 1-\left(x^2+x+\dfrac{1}{4}\right)+\dfrac{1}{4} = \dfrac{5}{4} - \left(x+\dfrac{1}{2}\right)^2$

(c) $5-(x^2-5x) = 5-\left(x^2-5x+\dfrac{25}{4}\right)+\dfrac{25}{4} = \dfrac{45}{4} - \left(x-\dfrac{5}{2}\right)^2$

(d) $1-(x^2+3x) = 1-\left(x^2+3x+\dfrac{9}{4}\right)+\dfrac{9}{4} = \dfrac{13}{4} - \left(x+\dfrac{3}{2}\right)^2$

(e) $-10-\left(x^2-x\right)=-10-\left(x^2-x+\dfrac{1}{4}\right)+\dfrac{1}{4}=-\dfrac{39}{4}-\left(x-\dfrac{1}{2}\right)^2$

(f) $-8-\left(x^2-7x\right)=-8-\left(x^2-7x+\dfrac{49}{4}\right)+\dfrac{49}{4}=\dfrac{17}{4}-\left(x-\dfrac{7}{2}\right)^2$

(g) $21-\left(x^2+7x\right)=21-\left(x^2+7x+\dfrac{49}{4}\right)+\dfrac{49}{4}=\dfrac{133}{4}-\left(x+\dfrac{7}{2}\right)^2$

Checks may be carried out as for questions 1 & 2 and are omitted here.

5. $y=x^2-10x+12=\left(x^2-10x+25\right)-25+12=(x-5)^2-13$. Therefore, the vertex has coordinates $(5,\ -13)$. As the axis of symmetry passes through the vertex, it has equation $x=5$. Alternatively, the equation of the axis of symmetry is $x=-\dfrac{b}{2a}=-\dfrac{-10}{2\times 1}=5$.

6. Let $f(x)=ax^2+bx+c$. As the parabola passes through $(0,\ 6)$, $f(0)=6\Rightarrow c=6$. Since the axis of symmetry is $x=\dfrac{5}{2}$, $-\dfrac{b}{2a}=\dfrac{5}{2}\Rightarrow 5a=-b$ and, as the parabola passes through $(3,\ 0)$, $f(3)=0\Rightarrow 9a+3b+6=0$. Therefore, $3a+b+2=0$ and $b=-5a\Rightarrow a=1$. So $b=-5$ and $f(x)=x^2-5x+6$. Alternatively, if the parabola passes through $(3,\ 0)$ and has axis of symmetry $x=\dfrac{5}{2}$, then, by symmetry, the parabola also passes through $=(2,\ 0)$. Therefore, $f(x)=(x-3)(x-2)=x^2-5x+6$.

Exercise 2.5

1. $2\left(x^2-2x\right)+1=2\left(x^2-2x+1\right)-1\times 2+1=2(x-1)^2-1$

2. $4\left(x^2+\dfrac{3}{2}x\right)-9=4\left(x^2+\dfrac{3}{2}x+\dfrac{9}{16}\right)-\dfrac{9}{16}\times 4-9=4\left(x+\dfrac{3}{4}\right)^2-\dfrac{45}{4}$

3. $11-2\left(x^2+2x\right)=11-2\left(x^2+2x+1\right)+1\times 2=11-2(x+1)^2+2=13-2(x+1)^2$

4. $3\left(x^2-\dfrac{10}{3}x\right)+15=3\left(x^2-\dfrac{10}{3}x+\dfrac{25}{9}\right)-\dfrac{25}{9}\times 3+15=3\left(x-\dfrac{5}{3}\right)^2+\dfrac{20}{3}$

5. $1 - 4\left(x^2 + \frac{1}{2}x\right) = 1 - 4\left(x^2 + \frac{1}{2}x + \frac{1}{16}\right) - \frac{1}{16} \times 4 = \frac{5}{4} - 4\left(x + \frac{1}{4}\right)^2$

6. $-5 - 2\left(x^2 - 4x\right) = -5 - 2\left(x^2 - 4x + 4\right) + 8 = 3 - 2\left(x - 2\right)^2$

Exercise 2.6

1. (a)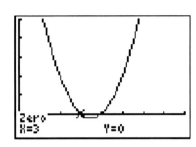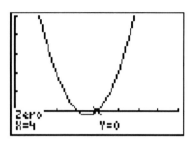

   Intercepts are 3 and 4 therefore $x^2 - 7x + 12 = (x-3)(x-4)$

   A similar method may be used for the other parts of the question. Only the factored solutions are shown.

   (b) $(x+3)(x+2)$ (c) $(x+9)(x-4)$ (d) $(x-10)(x+9)$ (e) $(x-1)^2$

   (f) $(5+x)(3-x)$ (g) $(3+x)(2-x)$ (h) $(7-x)(1+x)$ (i) $(x-5)(x+2)$

   (j) $(5+x)(1-x)$ (k) $(x+11)(x-3)$ (l) $(x+7)(x+4)$

2. (a) $x^2 - 3x - 4x + 12 = x(x-3) - 4(x-3) = (x-3)(x-4)$

   (b) $x^2 + 3x + 2x + 6 = x(x+3) + 2(x+3) = (x+3)(x+2)$

   (c) $x^2 + 9x - 4x - 36 = x(x+9) - 4(x+9) = (x+9)(x-4)$

   (d) $x^2 - 10x + 9x - 90 = x(x-10) + 9(x-10) = (x-10)(x+9)$

   (e) $x^2 - x - x + 1 = x(x-1) - 1(x-1) = (x-1)(x-1) = (x-1)^2$

   (f) $15 - 5x + 3x - x^2 = 5(3-x) + x(3-x) = (3-x)(5+x)$

   (g) $6 - 3x + 2x - x^2 = 3(2-x) + x(2-x) = (2-x)(3+x)$

(h) $7+7x-x-x^2 = 7(1+x)-x(1+x) = (1+x)(7-x)$

(i) $x^2 -5x+2x-10 = x(x-5)+2(x-5) = (x-5)(x+2)$

(j) $5-5x+x-x^2 = 5(1-x)+x(1-x) = (1-x)(5+x)$

(k) $x^2 +11x-3x-33 = x(x+11)-3(x+11) = (x+11)(x-3)$

(l) $x^2 +4x+7x+28 = x(x+4)+7(x+4) = (x+4)(x+7)$

Exercise 2.7

1. (a) $x^2 +10x-2x-20 = 0 \Rightarrow x(x+10)-2(x+10)=0 \Rightarrow (x+10)(x-2)=0 \Rightarrow x=-10,\ x=2$

   (b) $x\left(x-\dfrac{72}{x}\right) = x \Rightarrow x^2 -72 = x \Rightarrow x^2 -x-72 = 0$
   $\Rightarrow x^2 -9x+8x-72 = 0 \Rightarrow x(x-9)+8(x-9)=0 \Rightarrow (x-9)(x+8)=0 \Rightarrow x=9,\ x=-8$

   (c) $3x-x^2 = 9x+9 \Rightarrow x^2+6x+9=0 \Rightarrow x^2+3x+3x+9$
   $\Rightarrow x(x+3)+3(x+3)=0 \Rightarrow (x+3)(x+3)=(x+3)^2 = 0 \Rightarrow x=-3$

   (d) $x^2 -8x+16 = 8x-47 \Rightarrow x^2 -16x+63 = 0 \Rightarrow x^2 -9x-7x+63 = 0$
   $\Rightarrow x(x-9)-7(x-9)=0 \Rightarrow (x-9)(x-7)=0 \Rightarrow x=9,\ x=7$

2. (a) $x = \dfrac{-8 \pm \sqrt{64+12}}{6} = -\dfrac{4}{3} \pm \dfrac{\sqrt{76}}{6} = -\dfrac{4}{3} \pm \dfrac{\sqrt{19}}{3}$, so $p = -\dfrac{4}{3},\ k = \pm\dfrac{1}{3},\ q=19$.

   (b) $x = \dfrac{-7 \pm \sqrt{49-44}}{2} = -\dfrac{7}{2} \pm \dfrac{1}{2}\sqrt{5}$, so $p = -\dfrac{7}{2},\ k = \pm\dfrac{1}{2},\ q=5$.

   (c) $x = \dfrac{3 \pm \sqrt{9+20}}{10} = \dfrac{3}{10} \pm \dfrac{1}{10}\sqrt{29}$, so $p = \dfrac{3}{10},\ k = \pm\dfrac{1}{10},\ q=29$.

   (d) $x = \dfrac{17 \pm \sqrt{289-272}}{4} = \dfrac{17}{4} \pm \dfrac{1}{4}\sqrt{17}$, so $p = \dfrac{17}{4},\ k = \pm\dfrac{1}{4},\ q=17$.

3. (a)

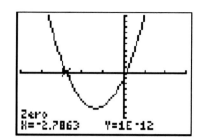 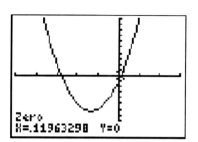

Therefore, $x = -2.7863$, $x = 0.1196$. As $x = -\frac{4}{3} \pm \frac{1}{3}\sqrt{19} = -2.786299648$, 0.1196329812, the calculator solutions agree with the algebraic solutions up to four decimal places.

The other parts may be checked in a similar way.

4. (a) $(x^2 - 4x + 4) - 4 - 11 = 0 \Rightarrow (x-2)^2 = 15$; therefore, $a = 2$, $b = 15$ and $x - 2 = \pm\sqrt{15}$
$\Rightarrow x = 2 + \sqrt{15}$, $x = 2 - \sqrt{15}$.

(b) $(x^2 - 10x + 25) - 25 + 23 = 0 \Rightarrow (x-5)^2 - 2 = 0 \Rightarrow x - 5 = \pm\sqrt{2}$. So $x = 5 - \sqrt{2}$, $x = 5 + \sqrt{2}$

Exercise 2.8

1. (a) $\Delta = 8$ (b) $\Delta = 0$, equal solutions. (c) $\Delta = -4$, no real solutions.

   (d) $\Delta = -84$, no real solutions. (e) $\Delta = 13$ (f) $\Delta = 49$

   (g) $\Delta = 29$ (h) $\Delta = 37$ (i) $\Delta = 0$, equal solutions.

2. For no real solutions, $\Delta < 0$ and as $\Delta = (-1)^2 - 4k(-1) = 1 + 4k$, $1 + 4k < 0 \Rightarrow k < -\frac{1}{4}$.

3. For equal solutions $\Delta = 0$ and as $\Delta = 4^2 - 4k(k+3) = 16 - 12k - 4k^2$, $16 - 12k - 4k^2 = 0$
$\Rightarrow k^2 + 3k - 4 = 0 \Rightarrow k^2 + 4k - k - 4 = 0 \Rightarrow k(k+4) - 1(k+4) = 0 \Rightarrow (k+4)(k-1) = 0$
$\Rightarrow k = -4, 1$.

4. If $y = kx - k$ and $y = x^2 + 5x + 10$ touch then $kx - k = x^2 + 5x + 10$ has equal solutions.
Therefore $x^2 + (5-k)x + (10+k) = 0$ has $\Delta = 0$. But $\Delta = (5-k)^2 - 4(10+k)$
$= 25 - 10k + k^2 - 40 - 4k = k^2 - 14k - 15$ so that $k^2 - 14k - 15 = 0$
$\Rightarrow k(k-15) + 1(k-15) = 0 \Rightarrow (k-15)(k+1) = 0 \Rightarrow k = 15, k = -1$.

5. (a) If the curves touch, the solution of
$$x^2 + \frac{1}{8} = 3x - x^2 - 1 \Rightarrow 2x^2 - 3x + \frac{9}{8} = 0 \Rightarrow 16x^2 - 24x + 9 = 0$$
will have two equal solutions. Therefore, $\Delta = 0$ but $\Delta = 24^2 - 4 \times 19 \times 9 = 576 - 576 = 0$, and therefore the curves do touch each other. $16x^2 - 24x + 9 = 0 \Rightarrow (4x - 3)^2 = 0$
$\Rightarrow x = \frac{3}{4} \Rightarrow y = \frac{11}{16}$. Therefore the coordinates of the point of contact are $\left(\frac{3}{4}, \frac{11}{16}\right)$.

(b) If the curves touch, there will be two equal solutions of $x^2 + kx + k = -2 - 2x - x^2$
$\Rightarrow 2x^2 + (2 + k)x + (k + 2) = 0$. This equation will have equal solutions if $\Delta = 0$ but
$\Delta = (2 + k)^2 - 8(k + 2) = (k + 2)((k + 2) - 8) = (k + 2)(k - 6)$.
Therefore, $(k + 2)(k - 6) = 0 \Rightarrow k = -2, 6$.
For $k = -2$, $2x^2 = 0 \Rightarrow x = 0 \Rightarrow y = -2$ and point of contact is $(0, -2)$.
For $k = 6$, $2x^2 + 8x + 8 = 0 \Rightarrow x^2 + 4x + 4 = 0 \Rightarrow (x + 2)^2 = 0 \Rightarrow x = -2 \Rightarrow y = -2$ and point of contact is $(-2, -2)$.

Exercise 2.9

1. (a) $(f \circ g)(1) = f(g(1)) = f(3) = 16$ (b) $(g \circ f)(4) = g(f(4)) = g(19) = -15$

   (c) $(f \circ f)(2) = f(f(2)) = f(13) = 46$ (d) $(g \circ g)(-1) = g(g(-1)) = g(5) = -1$

2. (a) $(g \circ f)(x) = g(f(x)) = g(x + 1) = 3(x + 1) = 3x + 3$

   (b) $(f \circ g)(x) = f(g(x)) = f(3x) = 3x + 1$
   $f(x) \times g(x) = (x + 1) \times (3x) = 3x^2 + 3x \neq 3x + 1 = (f \circ g)(x)$

   (c) $(f \circ f)(x) = f(f(x)) = f(x + 1) = (x + 1) + 1 = x + 2$

3. (a) $(g \circ f)(x) = \frac{5}{2}x + \frac{7}{2}$ (b) $(f \circ g)(x) = \frac{5}{2}x + 19$

   (c) $(g \circ h)(x) = \frac{9}{2} - x$ (d) $(f \circ h)(x) = 4 - 10x$

4. (a) $f \circ g : x \mapsto 11 - 3x^2$ (b) $g \circ h : x \mapsto \frac{1}{x^2} - 3, x \neq 0$ (c) $h \circ f : x \mapsto \frac{1}{2 - 3x}, x \neq \frac{2}{3}$

   (d) $h \circ h = h^2 : x \mapsto x$ (or $h^2 = i$) (e) $g \circ g = g^2 : x \mapsto (x^2 - 3)^2 - 3 = x^4 - 6x^2 + 6$

5. $(g \circ h)(x) = (1-x)+5 = 6-x \Rightarrow (f \circ (g \circ h)) = f(6-x) = 2(6-x)$
   $(f \circ g)(x) = f(x+5) = 2(x+5) \Rightarrow ((f \circ g) \circ h)(x) = (f \circ g)(h(x)) = 2(h(x)+5)$
   $= 2((1-x)+5) = 2(6-x)$. Therefore, $f \circ (g \circ h) = (f \circ g) \circ h$.

6. (a) $(g \circ f)(x) = g(f(x)) = g(2x-1) = \frac{1}{2}(2x-1) + \frac{1}{2} = x - \frac{1}{2} + \frac{1}{2} = x$.
   Therefore, $(g \circ f)(x) = x = i(x) \Rightarrow g \circ f = i$.

   (b) $f^2(x) = (f \circ f)(x) = f(f(x)) = f(2-x) = 2-(2-x) = x$. So $f^2(x) = x = i(x) \Rightarrow f^2 = i$.

7. (a) $f^{-1}(x) = \frac{x-1}{3}$ (b) $g^{-1}(x) = 3-x$ (c) $h^{-1}(x) = \frac{1-x}{4}$ (d) $f^{-1}(x) = \frac{3}{x}, x \neq 0$

   (e) $g^{-1}(x) = \frac{3x-1}{2}$ (f) $h^{-1}(x) = \frac{2(x+2)}{x-1}, x \neq 1$

8. (a) $f^{-1}(4) = -1$ (b) $g^{-1}(-1) = -\frac{1}{4}$ (c) $h^{-1}(2) = 0$

   (d) $(f \circ g)(x) = 4x+5 \Rightarrow (f \circ g)^{-1}(x) = \frac{x-5}{4} \Rightarrow (f \circ g)^{-1}(3) = -\frac{1}{2}$

   (e) $(h^{-1} \circ g^{-1})(x) = h^{-1}\left(\frac{x}{4}\right) = 2 - \frac{x}{4} \Rightarrow (h^{-1} \circ g^{-1})(-1) = \frac{9}{4}$

9. $(f \circ g)(x) = 5(1-2x) = 5-10x \Rightarrow x = 5-10y \Rightarrow 10y = 5-x$. Therefore, $(f \circ g)^{-1}(x) = \frac{5-x}{10}$.
   $f^{-1}(x) = \frac{x}{5}, g^{-1}(x) = \frac{1-x}{2} \Rightarrow (g^{-1} \circ f^{-1})(x) = g^{-1}\left(\frac{x}{5}\right) = \frac{1-\frac{x}{5}}{2} = \frac{5-x}{10}$.
   Therefore, $(f \circ g)^{-1} = g^{-1} \circ f^{-1}$.

Exercise 2.10

1. (a)   (b)   (c)

(d)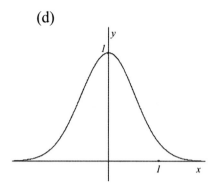

(e)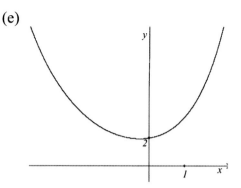

2. The graph of $y = \log_2 x$ is shown by the thick curve.

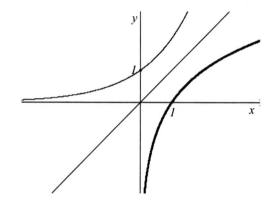

3.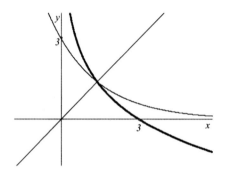

The graph of $y = 2(1 - \log_3 x)$ is shown by the thick line.

4. (a) 5 (b) $a^{3\log_a 2} = a^{\log_a 2^3} = a^{\log_a 8} = 8$ (c) $a^{-2\log_a 3} = a^{\log_a 3^{-2}} = a^{\log_a \frac{1}{9}} = \frac{1}{9}$

(d) $a^{\frac{1}{2}\log_a 16} = a^{\log_a 16^{\frac{1}{2}}} = a^{\log_a 4} = 4$ (e) $a^{\log_{a^2} 3} = a^{\frac{\log_a 3}{\log_a a^2}} = a^{\frac{1}{2}\log_a 3} = a^{\log_a \sqrt{3}} = \sqrt{3}$

5. (a) $\log_3 a = \dfrac{\log_a a}{\log_a 3} = \dfrac{1}{4.931} = 0.203$ (b) $\log_a 27 = 3\log_a 3 = 3 \times 4.931 = 14.8$

(c) $\log_{a} a^2 = \dfrac{\log_a a^2}{\log_a 3a} = \dfrac{2}{\log_a 3 + \log_a a} = \dfrac{2}{4.931 + 1} = 0.337$

6. (a) $x = \log_2 5 = \dfrac{\log_{10} 5}{\log_{10} 2} = 2.3219$

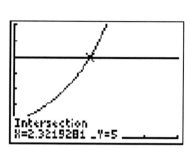

(b) $2x - 3 = \log_5 39 \Rightarrow x = \dfrac{1}{2}(\log_5 39 + 3)$
$= \dfrac{1}{2}\left(\dfrac{\log_{10} 39}{\log_{10} 5} + 3\right) = 2.6381$

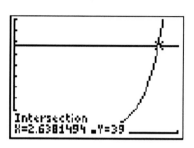

(c) $3^x = 243 = 3^5 \Rightarrow x = 5$

(d) $1 - x = \log_3 4 = \dfrac{\log_{10} 4}{\log_{10} 3} \Rightarrow x = 1 - \dfrac{\log_{10} 4}{\log_{10} 3} = -0.2619$

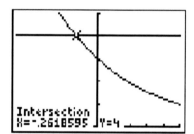

(e) $2^{3-x} = \dfrac{1}{16} = 2^{-4} \Rightarrow 3 - x = -4 \Rightarrow x = 7$

Exercise 2.11

1. $y = e^{3x}$. To find $f^{-1}(x)$ exchange x and y, so
$x = e^{3y} \Rightarrow \ln x = 3y \Rightarrow y = \dfrac{1}{3}\ln x$ and $f^{-1}(x) = \dfrac{1}{3}\ln x,\ x > 0$.

The transformation is a reflection in the line $y = x$.

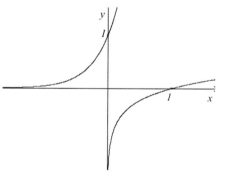

2. (a) $y = e^{\frac{1}{x}}$. To find $g^{-1}(x)$ exchange x and y so $x = e^{\frac{1}{y}} \Rightarrow \ln x = \dfrac{1}{y} \Rightarrow y = \dfrac{1}{\ln x}$.

Therefore, $g^{-1}(x) = \dfrac{1}{\ln x},\ x > 0$.

(b) $g(x)$ is shown with a thin line and $g^{-1}(x)$ is shown with a thick line. The transformation is a reflection in the line $y = x$.

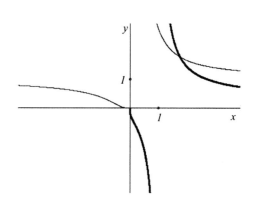

3. (a) $y = 3e^{-x} \Rightarrow x = 3e^{-y} \Rightarrow -y = \ln\dfrac{x}{3} \Rightarrow y = \ln\dfrac{3}{x}$. Therefore, $f^{-1} : x \mapsto \ln\dfrac{3}{x}$, $x > 0$.

(b) $y = 2\ln(x+1) \Rightarrow x = 2\ln(y+1) \Rightarrow y+1 = e^{\frac{x}{2}} \Rightarrow y = e^{\frac{x}{2}} - 1$. Therefore, $f^{-1} : x \mapsto e^{\frac{x}{2}} - 1$.

(c) $y = e^{-x^2} \Rightarrow x = e^{-y^2} \Rightarrow -y^2 = \ln x \Rightarrow y = \sqrt{-\ln x} \Rightarrow y = \sqrt{\ln\dfrac{1}{x}}$.

Therefore, $f^{-1} : x \mapsto \sqrt{\ln\dfrac{1}{x}}$, $x > 0$.

(d) $y = \ln(3x-4) \Rightarrow x = \ln(3y-4) \Rightarrow 3y - 4 = e^x \Rightarrow 3y = e^x + 4 \Rightarrow y = \dfrac{1}{3}(e^x + 4)$.

Therefore, $f^{-1} : x \mapsto \dfrac{1}{3}(e^x + 4)$.

4. (a) $3^x = e^{\ln 3^x} = e^{x\ln 3} = e^{1.10x}$

(b) $4^{-2x} = e^{\ln 4^{-2x}} = e^{-2x\ln 4} = e^{-2.77x}$

(c) $10^{0.1x} = e^{\ln 10^{0.1x}} = e^{0.1x\ln 10} = e^{0.230x}$

(d) $7^{0.5x} = e^{\ln 7^{0.5x}} = e^{0.5x\ln 7} = e^{0.973x}$

(e) $0.4^x = e^{\ln 0.4^x} = e^{x\ln 0.4} = e^{-0.916x}$

5. (a) $\ln 2^{x+1} = \ln 5 \Rightarrow (x+1)\ln 2 = \ln 5 \Rightarrow x + 1 = \dfrac{\ln 5}{\ln 2} \Rightarrow x = \dfrac{\ln 5}{\ln 2} - 1 = 1.3219$

(b) $\ln 3^{2x} = \ln 22 \Rightarrow 2x \ln 3 = \ln 22 \Rightarrow 2x = \dfrac{\ln 22}{\ln 3} \Rightarrow x = \dfrac{\ln 22}{2\ln 3} = 1.4068$

(c) $\ln 1.3^{-2x} = \ln 0.165 \Rightarrow -2x \ln 1.3 = \ln 0.165 \Rightarrow x = -\dfrac{\ln 0.165}{2\ln 1.3} = 3.4338$

(d) $4^{2x-3} = 2.4 \Rightarrow (2x-3)\ln 4 = \ln 2.4 \Rightarrow 2x - 3 = \dfrac{\ln 2.4}{\ln 4} \Rightarrow x = \dfrac{1}{2}\left(\dfrac{\ln 2.4}{\ln 4} + 3\right) = 1.8158$

Exercise 2.12

1. (a) $f(3) = 100e^{0.5 \times 3} = 100e^{1.5} \approx 448.168907$, so the population is 448 organisms.

   (b) $f(7) = 100e^{0.5 \times 7} = 100e^{3.5} \approx 3311.545196$, so the population is 3310 organisms.

   (c) $f(28) = 100e^{0.5 \times 28} = 100e^{14} \approx 1.202604.284 \times 10^8$, so the population is about 120 million organisms.

2. (a) $n(0) = 250$

   (b) Number of birds in January 2005 is $n(3) = 250e^{-0.24} = 197$. Number of birds in January 2006 is $n(4) = 250e^{-0.32} = 182$. Therefore, the decline during 2005 is $197 - 182 = 15$ birds.

   (c) You need to find t such that $250e^{-0.08t} = 100 \Rightarrow e^{-0.8t} = 0.4 \Rightarrow -0.08t = \ln 0.4$
   $= -0.91629 \Rightarrow t = 11.454$. Therefore, the population of birds falls below 100 after about 11.5 years, that is during the year 2013.

3. Jim's investment is worth, in dollars, $10000e^{0.043t}$ after t years. So after 5 years, it is worth $10000e^{0.043 \times 5} = 10000e^{0.215} = 12398.62$, and the interest earned during this time is $2398.62.

   (a) If interest is compounded yearly, Jim's investment is worth, in dollars,
   $10000(1.043)^5 - 10000 = 2343.02$. So Jim overestimates his interest by $55.60.

   (b) If interest is compounded monthly, Jim's investment is worth, in dollars,
   $10000\left(1 + \frac{4.3}{1200}\right)^{60} - 10000 = 10000(1.0035833)^{60} - 10000 = 2393.86$. So Jim overestimates his interest by $4.76.

4. (a) $n_A(0) = 5000,\ n_B(0) = 3500$ (b) $n_A(10) = 5000e^{0.02 \times 10} = 5000e^{0.2} = 6110$

   (c) $n_B(t) = 7000 \Rightarrow 3500e^{0.025t} = 7000 \Rightarrow e^{0.025t} = 2 \Rightarrow 0.025t = \ln 2 \Rightarrow t = 40\ln 2 = 27.7$.
   Therefore it will take 28 years for the number of B seals to double.

   (d) $n_A(t) = n_B(t) \Rightarrow 5000e^{0.02t} = 3500e^{.025t} \Rightarrow \frac{e^{0.02t}}{e^{0.025t}} = \frac{35}{50} = 0.7 \Rightarrow e^{(0.02 - 0.025)t} = 0.7$
   $\Rightarrow e^{-0.005t} = 0.7 \Rightarrow -0.005t = \ln 0.7 \Rightarrow -0.005t = -0.3566749 \Rightarrow t = 71.3$. Therefore, the number of seals will be equal after about 70 years.

5. After 3 weeks (21 days), the support for candidate A is estimated at $28e^{0.004 \times 21} = 28e^{0.084}$ = 30.5 %. Support for candidate B is estimated at $40e^{-0.012 \times 21} = 40e^{-0.252} = 31.1\%$. Therefore, on the basis of the survey, candidate B is (just) likely to win.

6. (a) $f(0) = 30 \Rightarrow n_0\left(1 - 0.9e^{-0.15 \times 0}\right) = n_0(1 - 0.9) = 0.1n_0 \Rightarrow n_0 = \dfrac{30}{0.1} = 300$

   (b)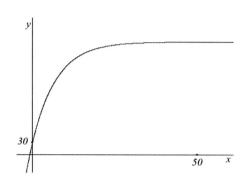

   (c) $f(6) = 300\left(1 - 0.9e^{-0.15 \times 6}\right) = 300\left(1 - 0.9e^{-0.9}\right) = 190$

   (d) The graph has a horizontal asymptote at $n = 300$, indicating that after a long time (that is for large values of t) the pond will have 300 trout.

   Alternatively: $f(20) = 300\left(1 - 0.9e^{-3}\right) = 286.557$,
   $f(50) = 300\left(1 - 0.9e^{-7.5}\right) = 299.851$
   $f(100) = 300\left(1 - 0.9e^{-15}\right) = 299.9999174$
   from which it is clear that, as t increases, n gets closer and closer to 300.

Solutions to Unit 3 Exercises

Exercise 3.1

1. (a) 120° (b) 72° (c) 150° (d) 80° (e) 540° (f) 54° (g) 15° (h) 70° (i) 142.5°

2. (a) $\dfrac{2\pi}{3}$ (b) $\dfrac{5\pi}{9}$ (c) $\dfrac{\pi}{10}$ (d) $\dfrac{\pi}{180}$ (e) $\dfrac{34\pi}{15}$ (f) $\dfrac{7\pi}{36}$ (g) $\dfrac{7\pi}{60}$ (h) $\dfrac{37\pi}{120}$ (i) $\dfrac{7\pi}{6}$

3. (a) 10cm (b) 1.75cm (c) 3.49cm (d) 12.0cm

4. (a) 0.0633 = 3.63° (b) 0.949 = 54.4° (c) 0.224 = 12.8° (d) 0.499 = 28.6°

5. (a) 28.3cm² (b) 18.8cm² (c) 23.6cm² (d) 23.2cm² (e) 5.65cm² (f) 22.9cm²

6. (a) 1.39 (b) 1.04 (c) 0.871

7. Diameter of the moon is approximately equal to the length of the arc of the sector. Arc length is 382100θ, where θ is the angle of the sector in radians. So $\theta = \dfrac{0.5167\pi}{180} = 0.0090181$ and the arc length is $382100 \times 0.0090181 = 3445.82$. Therefore, the diameter of the Moon is 3450km.

8. A_1 is the starting position of A. When the triangle is rotated about B, A_1 is rotated about B through $\dfrac{2\pi}{3}$ to A_2. The length of the arc through which A moves is $6 \times \dfrac{2\pi}{3} = 4\pi$ cm. When the

   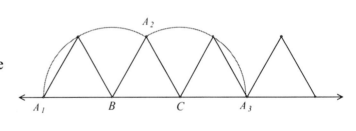

   triangle is then rotated about C, A_2 is rotated through an angle of $\dfrac{2\pi}{3}$ about C to A_3. This arc length is also 4π cm. Finally, the triangle is rotated about A, so obviously A does not move. Therefore, the path traveled by A is 8π cm ≈ 25.1cm.

9. (a) $\dfrac{\theta}{2\pi} = \dfrac{1}{50} \Rightarrow \theta = \dfrac{2\pi}{50} = 0.12566$. Therefore, the angle between sun's rays and stick is 0.126 radians.

   (b) $r\theta = 850 \Rightarrow r = \dfrac{850}{0.12566} = 6764.3$. Therefore radius of Earth is 6760km.

Exercise 3.2

1. (a) $\sin\left(\pi + \dfrac{\pi}{2}\right) = -\sin\dfrac{\pi}{2}$ (b) $\sin\left(\pi - \dfrac{3\pi}{8}\right) = \sin\dfrac{3\pi}{8}$ (c) $\sin\left(\pi + \dfrac{3\pi}{10}\right) = -\sin\dfrac{3\pi}{10}$

   (d) $\sin\left(2\pi - \dfrac{\pi}{4}\right) = -\sin\dfrac{\pi}{4}$ (e) $\sin(360° - 60°) = -\sin 60°$ (f) $\sin(180° - 45°) = \sin 45°$

   (g) $\sin(180° + 56°) = -\sin 56°$ (h) $\sin(360° - 23°) = -\sin 23°$

2. (a) $\cos\left(2\pi - \dfrac{\pi}{2}\right) = \cos\dfrac{\pi}{2} \;(= 0)$ (b) $\cos\left(\pi - \dfrac{\pi}{5}\right) = -\cos\dfrac{\pi}{5}$

   (c) $\cos\left(\pi + \dfrac{\pi}{6}\right) = -\cos\dfrac{\pi}{6}$ (d) $\cos\left(\pi - \dfrac{\pi}{6}\right) = -\cos\dfrac{\pi}{6}$

   (e) $\cos(180° - 70°) = -\cos 70°$ (f) $\cos(180° + 30°) = -\cos 30°$

   (g) $\cos(360° - 50°) = \cos 50°$ (h) $\cos(180° - 47°) = -\cos 47°$

3. (a) $\tan\left(\pi + \dfrac{\pi}{4}\right) = \tan\dfrac{\pi}{4}$ (b) $\tan\left(\pi + \dfrac{\pi}{3}\right) = \tan\dfrac{\pi}{3}$ (c) $\tan\left(\pi - \dfrac{\pi}{8}\right) = -\tan\dfrac{\pi}{8}$

   (d) $\tan\left(\pi + \dfrac{\pi}{6}\right) = \tan\dfrac{\pi}{6}$ (e) $\tan(180° - 50°) = -\tan 50°$ (f) $\tan(360° - 40°) = -\tan 40°$

   (g) $\tan(180° + 80°) = \tan 80°$ (h) $\tan(360° - 10°) = -\tan 10°$

4. (a) $\dfrac{1}{\sqrt{2}}$ (b) $\dfrac{1}{2}$ (c) $-\dfrac{\sqrt{3}}{2}$ (d) $-\dfrac{1}{\sqrt{2}}$ (e) $-\dfrac{1}{\sqrt{2}}$ (f) $-\dfrac{1}{2}$ (g) $\dfrac{\sqrt{3}}{2}$ (h) $\dfrac{1}{2}$

   Parts (i) to (l) can be done by evaluating the sine and cosine values of the angle and then using
   $\tan\theta = \dfrac{\sin\theta}{\cos\theta}$ (i) -1 (j) $-\sqrt{3}$ (k) $-\dfrac{1}{\sqrt{3}}$ (l) $\dfrac{1}{\sqrt{3}}$

5. The rotational symmetry (half-turn about the origin) of the graph of $y = \sin\theta$ shows that for all values of θ, $\sin(-\theta) = -\sin\theta$.

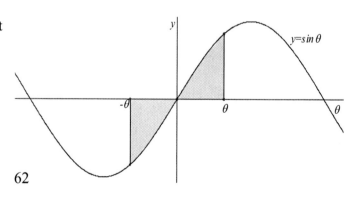

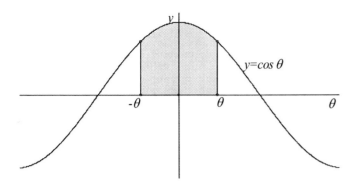

The reflectional symmetry (reflection in the y-axis) of the graph of $y = \cos\theta$ shows that for all values of θ, $\cos(-\theta) = \cos\theta$.

The rotational symmetry (half-turn about the origin) of the graph of $y = \tan\theta$ shows that for all values of θ, $\tan(-\theta) = -\tan\theta$.

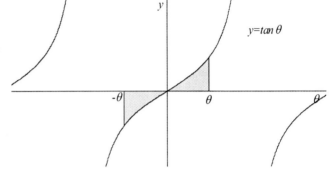

(a) $\sin\left(-\dfrac{\pi}{3}\right) = -\sin\dfrac{\pi}{3} = -\dfrac{\sqrt{3}}{2}$

(b) $\cos\left(-\dfrac{4\pi}{3}\right) = \cos\dfrac{4\pi}{3} = \cos\left(\pi + \dfrac{\pi}{3}\right) = -\cos\dfrac{\pi}{3} = -\dfrac{1}{2}$

(c) $\tan\left(-\dfrac{11\pi}{6}\right) = -\tan\left(\dfrac{11\pi}{6}\right) = \dfrac{1}{\sqrt{3}}$

Exercise 3.3

1. (a) $\dfrac{5}{8}$ (b) $x^2 + 5^2 = 8^2 \Rightarrow x = \sqrt{39}$ (c) $\cos\theta = \dfrac{\sqrt{39}}{8}$

   (d) $\cos^2\theta + \sin^2\theta = \left(\dfrac{\sqrt{39}}{8}\right)^2 + \left(\dfrac{5}{8}\right)^2 = \dfrac{39}{64} + \dfrac{25}{64} = \dfrac{39+25}{64} = 1$

2. (a) $\cos^2\theta = 1 - \sin^2\theta = 1 - \dfrac{4}{25} = \dfrac{21}{25} \Rightarrow \cos\theta = \pm\dfrac{\sqrt{21}}{5}$ but, as θ is acute $\cos\theta > 0$ so

   $\cos\theta = \dfrac{\sqrt{21}}{5}$.

   (b) As θ is obtuse $\cos\theta < 0$ so $\cos\theta = -\dfrac{\sqrt{21}}{5}$.

3. (a) $\sin^2\theta = 1 - \cos^2\theta \Rightarrow \sin^2\theta = 1 - \dfrac{4}{9} = \dfrac{5}{9} \Rightarrow \sin\theta = \pm\dfrac{\sqrt{5}}{3}$ but as $0 < \theta < \dfrac{\pi}{2}$, $\sin\theta = \dfrac{\sqrt{5}}{3}$ and $k = 5$.

   (b) $\sin^2\theta = 1 - \dfrac{81}{1681} = \dfrac{1600}{1681} \Rightarrow \sin\theta = \pm\dfrac{40}{41}$ but as $\pi < \theta < 2\pi$ so $\sin\theta < 0$ and $\sin\theta = -\dfrac{40}{41}$.

4. (a) $\sin^2\theta = 1 - \dfrac{25}{169} = \dfrac{144}{169} \Rightarrow \sin\theta = \pm\dfrac{12}{13}$. As $0 < \theta < \pi$, $\sin\theta > 0$, so $\sin\theta = \dfrac{12}{13}$.

   (b) $\sin 2\theta = 2\cos\theta\sin\theta = 2\left(-\dfrac{5}{13}\right)\left(\dfrac{12}{13}\right) = -\dfrac{120}{169}$

   (c) $\cos 2\theta = 2\cos^2\theta - 1 = 2\left(\dfrac{25}{169}\right) - 1 = -\dfrac{119}{169}$

5. $\cos^2\theta = 1 - \dfrac{9}{25} = \dfrac{16}{25} \Rightarrow \cos\theta = \pm\dfrac{4}{5}$, $\cos 2\theta = 2\cos^2\theta - 1 = 2\left(\dfrac{16}{25}\right) - 1 = \dfrac{7}{25}$

6. (a) $\cos^2\theta = 1 - \dfrac{49}{625} = \dfrac{576}{625} \Rightarrow \cos\theta = \pm\dfrac{24}{25}$

   (b) $\tan\theta = \dfrac{\sin\theta}{\cos\theta} = \dfrac{7/25}{\pm 24/25} = \pm\dfrac{7}{24}$

   (c) $\cos 2\theta = 1 - 2\sin^2\theta = 1 - 2\left(\dfrac{49}{625}\right) = \dfrac{527}{625}$

   (d) $\sin 2\theta = 2\cos\theta\sin\theta = 2\left(\pm\dfrac{24}{25}\right)\left(\dfrac{7}{25}\right) = \pm\dfrac{336}{625}$

7. (a) $1 + \tan\theta = 1 + \dfrac{\sin\theta}{\cos\theta} = \dfrac{\cos\theta + \sin\theta}{\cos\theta}$

   (b) $(\cos\theta + \sin\theta)^2 = \cos^2\theta + 2\cos\theta\sin\theta + \sin^2\theta = (\cos^2\theta + \sin^2\theta) + 2\cos\theta\sin\theta$
   $= 1 + \sin 2\theta$

   (c) $1 - \tan^2\theta = 1 - \dfrac{\sin^2\theta}{\cos^2\theta} = \dfrac{\cos^2\theta - \sin^2\theta}{\cos^2\theta} = \dfrac{\cos^2\theta - \sin^2\theta}{1 - \sin^2\theta} = \dfrac{(\cos\theta - \sin\theta)(\cos\theta + \sin\theta)}{(1 - \sin\theta)(1 + \sin\theta)}$

   (d) $\cos 2A = 1 - 2\sin^2 A$. If $A = 3\theta$, then $\cos 6\theta = 1 - 2\sin^2 3\theta \Rightarrow 2\sin^2 3\theta = 1 - \cos 6\theta$

   (e) $1 + \cos\theta + \cos 2\theta = 1 + \cos\theta + 2\cos^2\theta - 1 = \cos\theta + 2\cos^2\theta = \cos\theta(1 + 2\cos\theta)$

Exercise 3.4

1. (a) A stretch of factor 5, parallel to the y-axis.

   (b) A stretch of factor 3 parallel to the x-axis.

   (c) A translation of $\frac{\pi}{6}$ parallel to the positive x-axis.

2. (a) A stretch of factor $\frac{2}{5}$ parallel to the y-axis.

   (b) A translation of π parallel to the negative x-axis.

   (c) A stretch of factor 4 parallel to the x-axis.

3. (a) A stretch of factor $\frac{1}{2}$ parallel to the x-axis. A translation of $+1$ parallel to the y-axis.

   (b) A stretch of factor $\frac{1}{3}$ parallel to the x-axis. A stretch of factor 2 parallel to the y-axis.

   (c) A translation of $\frac{\pi}{2}$ parallel to the positive x-axis. A stretch of factor 2 parallel to the y-axis.

4. (a) 4π (b) π (c) π (d) $\frac{\pi}{3}$ (e) 6

   (f) $\cos 2x = 1 - 2\sin^2 x \Rightarrow \sin^2 x = \frac{1}{2} - \frac{1}{2}\cos 2x$. Therefore the period is π.

   (g) Use your calculator to draw the graph of $f(x) = \cos x + \sin x$, measure the period $= 2\pi$.

5. (a) $y = 2\sin x$ (b) $y = 1 + \cos x$ (c) $y = \tan\left(x - \frac{\pi}{3}\right)$

   (d) $y = \sin\frac{x}{2}$ (e) $y = \tan 3x$ (f) $y = -\cos x$

6. (a) Period is $\frac{2\pi}{3}$. (b) (i) 5 (ii) -5 (c) (i) $f(0) = -\frac{5\sqrt{3}}{2}$ (ii) $f\left(\frac{\pi}{2}\right) = \frac{5}{2}$

7. (a) Period is $\dfrac{2\pi}{120\pi} = \dfrac{1}{60}$ seconds (b) 60 (c) (i) 0 (ii) 4.52

8. (a) $t(1) = -1.96$. Therefore sunset is 1h 57 minutes before 18:00 which is 16:03.

  (b) $t(46) = -1.11$. Therefore sunset is 1h 7minutes before 18:00 which is 16:53.

  (c) $t(176) = 1.99$. Therefore sunset is 2h 0minutes after 18:00 which is 20:00.

Exercise 3.5

1. (a) $x = \dfrac{\pi}{6}, \dfrac{5\pi}{6}$ (b) $x = \dfrac{\pi}{4}, \dfrac{5\pi}{4}$ (c) $x = 0$ (d) $x = 0, \pi$ (e) $x = \dfrac{3\pi}{4}, \dfrac{5\pi}{4}$

  (f) $x = \dfrac{5\pi}{6}, \dfrac{11\pi}{6}$ (g) $x = \dfrac{2\pi}{3}, \dfrac{4\pi}{3}$ (h) $x = \dfrac{\pi}{4}, \dfrac{3\pi}{4}$ (i) $x = 0, \pi$

2. (a) $x = \dfrac{\pi}{2}, -\dfrac{\pi}{2}$ (b) $x = \dfrac{3\pi}{4}, -\dfrac{\pi}{4}$ (c) $x = -\dfrac{\pi}{6}, -\dfrac{5\pi}{6}$ (d) $x = -\dfrac{\pi}{6}, \dfrac{\pi}{6}$

  (e) $x = -\dfrac{\pi}{2}$ (f) $x = \dfrac{\pi}{3}, -\dfrac{2\pi}{3}$ (g) $x = -\dfrac{\pi}{4}, \dfrac{\pi}{4}$ (h) $x = \dfrac{\pi}{6}, \dfrac{5\pi}{6}$

  (i) $\sin\dfrac{2\pi}{3} = \dfrac{\sqrt{3}}{2} \Rightarrow x = -\dfrac{\pi}{6}, \dfrac{\pi}{6}$

3. (a) $\sin\left(x + \dfrac{\pi}{3}\right) = 1 \Rightarrow x + \dfrac{\pi}{3} = \dfrac{\pi}{2} \Rightarrow x = \dfrac{\pi}{6}$

  (b) $\tan\left(x - \dfrac{\pi}{2}\right) = 1 \Rightarrow x - \dfrac{\pi}{2} = \dfrac{\pi}{4} \Rightarrow x = \dfrac{3\pi}{4}$, $x - \dfrac{\pi}{2} = \dfrac{5\pi}{4} \Rightarrow x = \dfrac{7\pi}{4}$

  (c) $\sin x = -\dfrac{1}{2} \Rightarrow x = \dfrac{7\pi}{6}, \dfrac{11\pi}{6}$ (d) $\cos x = \dfrac{1}{\sqrt{2}} \Rightarrow x = \dfrac{\pi}{4}, \dfrac{7\pi}{4}$

  (e) $\sin^2 x = \dfrac{1}{2} \Rightarrow \sin x = \pm\dfrac{1}{\sqrt{2}} \Rightarrow x = \dfrac{\pi}{4}, \dfrac{3\pi}{4}, \dfrac{5\pi}{4}, \dfrac{7\pi}{4}$

  (f) $\cos^2 x = \dfrac{1}{4} \Rightarrow \cos x = \pm\dfrac{1}{2} \Rightarrow x = \dfrac{\pi}{3}, \dfrac{2\pi}{3}, \dfrac{4\pi}{3}, \dfrac{5\pi}{3}$

(g) $\tan^2 x = \dfrac{1}{3} \Rightarrow \tan x = \pm\dfrac{1}{\sqrt{3}} \Rightarrow x = \dfrac{\pi}{6}, \dfrac{5\pi}{6}, \dfrac{7\pi}{6}, \dfrac{11\pi}{6}$

4. (a) $\cos x = \dfrac{1}{\sqrt{2}} \Rightarrow x = \dfrac{\pi}{4}, \dfrac{7\pi}{4}, \quad \sin x = -\dfrac{1}{\sqrt{2}} \Rightarrow x = \dfrac{5\pi}{4}, \dfrac{7\pi}{4}.$ So $\tan x = -1 \Rightarrow x = \dfrac{7\pi}{4}$

(b) $\cos 2x = \dfrac{1}{2} \Rightarrow 2\cos^2 x - 1 = \dfrac{1}{2} \Rightarrow \cos^2 x = \dfrac{3}{4} \Rightarrow \cos x = \pm\dfrac{\sqrt{3}}{2} \Rightarrow x = \pm\dfrac{\pi}{6}$

Exercise 3.6

1. (a) $\sin\theta(\sin\theta + 1) = 0 \Rightarrow \sin\theta = 0, \ \sin\theta = -1 \Rightarrow \theta = 0, \pi, \dfrac{3\pi}{2}$

(b) $2\cos^2\theta + \cos\theta - 1 = 0 \Rightarrow 2\cos^2\theta + 2\cos\theta - \cos\theta - 1 = 0$
$\Rightarrow 2\cos\theta(\cos\theta + 1) - 1(\cos\theta + 1) = 0 \Rightarrow (\cos\theta + 1)(2\cos\theta - 1) = 0$
$\Rightarrow \cos\theta = -1 \text{ and } \cos\theta = \dfrac{1}{2} \Rightarrow \theta = \dfrac{\pi}{3}, \pi, \dfrac{5\pi}{3}$

(c) $2\sin^2\theta = 1 \Rightarrow \sin^2\theta = \dfrac{1}{2} \Rightarrow \sin\theta = \pm\dfrac{1}{\sqrt{2}} \Rightarrow \theta = \dfrac{\pi}{4}, \dfrac{3\pi}{4}, \dfrac{5\pi}{4}, \dfrac{7\pi}{4}$

2. (a) $6\sin^2 x + 3\sin x - 2\sin x - 1 = 0 \Rightarrow 3\sin x(2\sin x + 1) - 1(2\sin x + 1) = 0$
$(2\sin x + 1)(\sin x - 1) = 0 \Rightarrow \sin x = -\dfrac{1}{2}, \ \sin x = \dfrac{1}{3} \Rightarrow x = -150°, -30°, 19.5°, 161°$

(b) $3\cos^2 x + 3\cos x - \cos x - 1 = 0 \Rightarrow 3\cos x(\cos x + 1) - 1(\cos x + 1) = 0$
$\Rightarrow (\cos x + 1)(3\cos x - 1) = 0 \Rightarrow \cos x = -1, \ \cos x = \dfrac{1}{3} \Rightarrow x = 70.5°, 180°, -70.5°$

(c) $4\tan^2 x - 4\tan x + 3\tan x - 3 = 0 \Rightarrow 4\tan x(\tan x - 1) + 3(\tan x - 1) = 0$
$\Rightarrow (\tan x - 1)(4\tan x + 3) = 0 \Rightarrow \tan x = 1, \ \tan x = -\dfrac{3}{4} \Rightarrow x = -135°, -36.9°, 45°\ 143°$

3. $2\sin^2 x = 1 - \cos x$ but $\cos^2 x + \sin^2 x = 1 \Rightarrow \sin^2 x = 1 - \cos^2 x$
$\Rightarrow 2(1 - \cos^2 x) = 1 - \cos x \Rightarrow 2 - 2\cos^2 x = 1 - \cos x \Rightarrow 2\cos^2 x - \cos x - 1 = 0$
$\Rightarrow 2\cos^2 x - 2\cos x + \cos x - 1 = 0 \Rightarrow 2\cos x(\cos x - 1) + 1(\cos x - 1) = 0$
$\Rightarrow (\cos x - 1)(2\cos x + 1) = 0 \Rightarrow \cos x = 1, \ \cos x = -\dfrac{1}{2} \Rightarrow x = -\dfrac{2\pi}{3}, 0, \dfrac{2\pi}{3}$

4. $\sin 2x = \cos x \Rightarrow 2\cos x \sin x = \cos x \Rightarrow 2\cos x \sin x - \cos x = 0$
$\Rightarrow \cos x(2\sin x - 1) = 0 \Rightarrow \cos x = 0, \sin x = \frac{1}{2}$. $\sin x = \frac{1}{2} \Rightarrow x = \frac{\pi}{6}$, as required, but also $x = \frac{5\pi}{6}$. When $\cos x = 0$, $x = \frac{\pi}{2}$.

5. (a) $\sin x = 4\cos x \Rightarrow \tan x = 4 \Rightarrow x = 76°, 256°$

   (b) $3\cos x + 2\sin x = 0 \Rightarrow \tan x = -\frac{3}{2} \Rightarrow x = 180° - 56.31°, 360° - 56.31° \Rightarrow x = 124°, 304°$

   (c) $\sin 2x = 3\cos^2 x \Rightarrow 2\sin x \cos x = 3\cos^2 x \Rightarrow 2\sin x \cos x - 3\cos^2 x = 0$
   $\Rightarrow \cos x(2\sin x - 3\sin x) = 0 \Rightarrow \cos x = 0$ and $2\sin x - 3\cos x = 0$
   $\Rightarrow \tan x = \frac{3}{2} \Rightarrow x = 56°, 90°, 236°, 270°$

6. As the graphs intersect in two places, there are two solutions in the required interval. The least value of x which satisfies this equation is found using the graphing calculator. $x = 75.2754°$, correct to four decimal places.

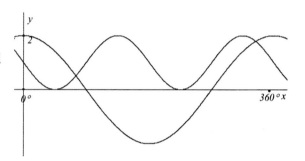

7. Method 1 is correct.

   In Method 2 $x = \frac{5\pi}{3} \Rightarrow \sin\frac{5\pi}{3} = -\frac{\sqrt{3}}{2}$; $\sqrt{3}\cos\frac{5\pi}{3} = \frac{\sqrt{3}}{2} \neq -\frac{\sqrt{3}}{2}$ and $x = \frac{2\pi}{3}$

   $\Rightarrow \sin\frac{2\pi}{3} = \frac{\sqrt{3}}{2}$; $\sqrt{3}\cos\frac{2\pi}{3} = -\frac{\sqrt{3}}{2} \neq \frac{\sqrt{3}}{2}$. By squaring both sides of the equation false solutions are introduced.

8. (a) $18:00 \Rightarrow t = 0 \Rightarrow \cos\left(\frac{2\pi}{365}(d+11)\right) = 0 \Rightarrow \frac{2\pi}{365}(d+11) = \frac{\pi}{2}, \frac{3\pi}{2}$
   $\Rightarrow d + 11 = \frac{365}{4} = 91.25, d + 11 = \frac{3 \times 365}{4} = 273.75 \Rightarrow d = 80.25, 262.75$
   Now $(31 + 28) + 21.25 = 80.25$ and $(31 + 28 + 31 + 30 + 31 + 30 + 31 + 31) + 19.75$.
   Therefore, the dates are March 21 and September 20.

   (b) $t = -1 \Rightarrow -2\cos\left(\frac{2\pi}{365}(d+11)\right) = -1 \Rightarrow \cos\left(\frac{2\pi}{365}(d+11)\right) = \frac{1}{2} \Rightarrow \frac{2\pi}{365}(d+11)$
   $= \frac{\pi}{3}, \frac{5\pi}{3} \Rightarrow d + 11 = 60.83, 304.17 \Rightarrow d = 49.83, 293.17$. Now $31 + 18.83 = 49.83$ and $(31 + 28 + 31 + 30 + 31 + 30 + 31 + 31 + 30) + 20.17$. Therefore, the dates are February 19 and October 20.

(c) An alternative, and perhaps quicker, approach is to use your calculator. 19:31 is equivalent to 1.51667h after 18:00 so $t = 1.51667$. This method could also have been used in parts (a) and (b)

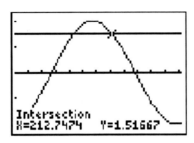

The diagram shows that $d = 212.75$ when $t = 1.51667$. The other value of d for $t = 1.51667$ can be found similarly and is $d = 130.25$. $(31 + 28 + 31 + 30) + 10.25 = 130.25$ and $(31 + 28 + 31 + 30 + 31 + 30 + 31) + 0.75 = 212.75$, and so the dates are May 10 and August 1.

9. (a) $h(0) = 16 \Rightarrow a = 16$, $a + b = 25 \Rightarrow b = 9$

(b) The period of the function is $\dfrac{25}{2}$ so that, $\dfrac{2\pi}{k} = \dfrac{25}{2} \Rightarrow k = \dfrac{4\pi}{25}$.

(c) $h(t) = 16 + 9\sin\dfrac{4\pi t}{25}$. $h(6) = 17.1$, $h(14) = 22.2$ so the depth of the water in the harbor at 06:00 is 17.1m and at 14:00 it is 22.2m.

(d) The depth of the water in the harbor is below 10m between 07.702 and 11.048, which is a time interval of 3.346h. However, there is another equal time interval later in the day so that the total time during which the depth is below 10m is 6.692h in a total time of 25 hours. Therefore, the proportion of time during which the depth of water is less than 10m is $\dfrac{6.692}{25} = 0.268 = 26.8\%$.

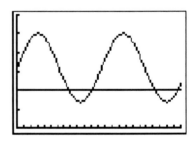

Exercise 3.7

1. (a) See diagram on right.

(b) Curve cuts x-axis in two places so there are two solutions.

(c) $x = 1.4221$

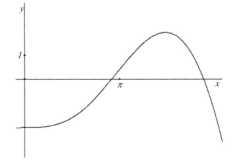

2. (a) 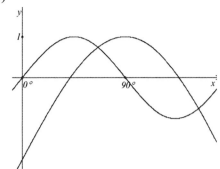 (b) $x = 66.1°, 155°$

3. (a) 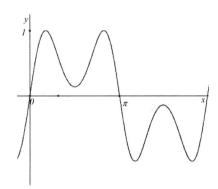 (b) (i) 2 (ii) 1 (iii) 4 (iv) 7

4. (a) 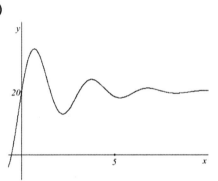 (b) Minimum clearance is 12.8cm, occurring when $t = 2.26$.

(c) The graph shows an enlarged portion of the function $h(t)$ and $y = 18$ for the given window.

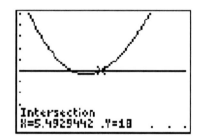

This shows that after 5.49 seconds the vibration is reduced to less than ± 2 cm.

Exercise 3.8

1. (a) $\hat{A} = 29.9°$, $\hat{B} = 56.3°$, $\hat{C} = 93.8°$ (b) $c = 13.9$, $\hat{A} = 78.5°$, $\hat{B} = 48.5°$

   (c) $a = 21.6$, $c = 22.3$, $\hat{C} = 79°$ (d) $\hat{A} = 35.7°$, $\hat{B} = 50.1°$, $\hat{C} = 94.1°$

   (e) $a = 6.00$, $b = 3.96$, $\hat{C} = 118°$ (f) $a = 2.36$, $\hat{B} = 47.6°$, $\hat{C} = 17.4°$

2. (a)

   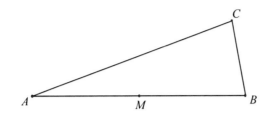

   (b) $BC = 2BN$, $N\hat{A}B = 10°$. Using sine rule $\sin 10° = \dfrac{NB}{15} \Rightarrow NB = 15\sin 10°$
   $\Rightarrow BC = 30\sin 10° = 5.21\text{cm}$

   (c) Consider $\triangle AMC$, $AM = 7.5\text{cm}$, $AC = 15\text{cm}$, $C\hat{A}M = 20°$
   $\Rightarrow CM^2 = 7.5^2 + 15^2 - 2 \times 7.5 \times 15 \cos 20° = 69.81916 \Rightarrow CM = 8.36\text{cm}$

3. $AB = AN + NB$. $\tan A\hat{C}N = \dfrac{AN}{5} \Rightarrow AN = 5\tan A\hat{C}N$, $\tan B\hat{C}N = \dfrac{NB}{5} \Rightarrow AN = 5\tan B\hat{C}N$, but $A\hat{C}N = 60°$, $N\hat{C}B = 40° \Rightarrow AB = 5\tan 60° + 5\tan 40° = 12.8558$. Therefore $AB = 12.9\text{m}$.

4. (a)

   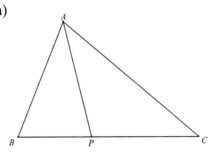

   (b) Using cosine rule, $AC^2 = 7.6^2 + 11.2^2 - 2 \times 7.6 \times 11.2 \cos 68° = 119.427 \Rightarrow AC = 10.9\text{cm}$.

   (c) $\cos A = (7.6^2 + 119.427 - 11.2^2)/(2 \times 7.6 \times 10.928) = 0.31153 \Rightarrow \hat{A} = 71.848°$
   Therefore $B\hat{A}P = 35.924° \Rightarrow A\hat{P}B = 76.076°$. Using sine rule, $\dfrac{AP}{\sin 68°} = \dfrac{7.6}{\sin 76.076°}$
   $\Rightarrow AP = \dfrac{7.6\sin 68°}{\sin 76.076°} = 7.2599 \Rightarrow AP = 7.26\text{cm}$.

(d) In $\triangle APC$ $\dfrac{PC}{\sin 35.92°} = \dfrac{7.2599}{\sin 40.152°} \Rightarrow PC = 6.61\text{cm}$

5. (a)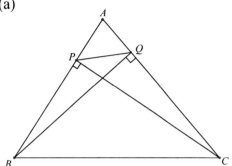

(b) Using cosine rule, $\cos A = \dfrac{(30^2 + 33^2 - 38^2)}{2 \times 30 \times 33} = 0.27525 \Rightarrow A = 74.0229°$ but $\cos A = \dfrac{AQ}{30}$

$\Rightarrow AQ = 30\cos 74.0229° = 8.25759$ and $\cos A = \dfrac{AP}{33} \Rightarrow AP = 33\cos 74.0229° = 9.08335$

Using cosine rule again,
$PQ^2 = 8.25759^2 + 9.08335^2 - 2 \times 8.25759 \times 9.08335 \cos 74.0229° \Rightarrow PQ = 10.5$

6. Proof of the cosine rule:
$a^2 = BN^2 + h^2$, $b^2 = (c + BN)^2 + h^2 \Rightarrow b^2 - a^2 = ((c + BN)^2 + h^2) - (BN^2 + h^2)$

$\Rightarrow b^2 - a^2 = c^2 + 2cBN$ but $\cos(180° - B) = \dfrac{BN}{a} \Rightarrow BN = a\cos(180° - B)$

$\Rightarrow b^2 - a^2 = c^2 + 2ca\cos(180° - B) \Rightarrow b^2 = c^2 + a^2 + 2ca\cos(180° - B)$ but as

$\cos(180° - B) = -\cos B$, so $b^2 = c^2 + a^2 - 2ca\cos B$.

Proof of the Sine Rule:
$\sin A = \dfrac{h}{b}$, $\sin(180° - B) = \dfrac{h}{a} \Rightarrow h = b\sin A$ and $h = a\sin(180° - B)$

$\Rightarrow a\sin(180° - B) = b\sin A$ but $\sin(180° - B) = \sin B \Rightarrow a\sin B = b\sin A \Rightarrow \dfrac{a}{\sin A} = \dfrac{b}{\sin B}$.

Exercise 3.9

1. (a) 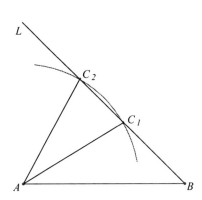   (b) $BC_1 = 5.5 \text{cm}, BC_2 = 10.3 \text{cm}$.

   (c) Using sine rule, $\dfrac{8}{\sin 44°} = \dfrac{11}{\sin C}$

   $\Rightarrow \sin C = \dfrac{11}{8}\sin 44° = 0.955155 \Rightarrow C = 72.776°, 107.224° \Rightarrow A = 63.224°, 28.776°$

   $\Rightarrow a = \dfrac{8\sin 63.224}{\sin 44°} = 10.282$, or $a = \dfrac{8\sin 28.776°}{\sin 44°} = 5.544$

   Therefore, $BC_1 = 5.54 \text{cm}$, $BC_2 = 10.3 \text{cm}$.

2. (a)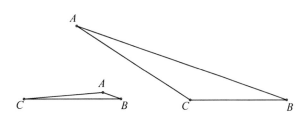

   (b) $27^2 = c^2 + 33^2 - 2 \times 33 \cos 19° \Rightarrow c^2 - 62.404c + 360 = 0 \Rightarrow c = 6.43, 56.0$

3. Using sine rule: $\dfrac{15}{\sin 11°} = \dfrac{51}{\sin R} \Rightarrow \sin R = \dfrac{51}{15}\sin 11° = 0.6487506 \Rightarrow R = 40.447°, 139.553°$.

   Therefore, the possible values of $\hat{R}$ are $40.4°$ and $140°$.

4. The figure shows the two possible positions of vertex B.

   $\dfrac{13.9}{\sin 33°} = \dfrac{16.6}{\sin B} \Rightarrow \sin B = \dfrac{16.6 \sin 33°}{13.9} = 0.650432$

   $\Rightarrow B = 40.574°, 139.426°$. For the large triangle:

   $\hat{C} = 180° - (33° + 40.574°) = 106.426°$.

   $c^2 = 16.6^2 + 13.9^2 - 2 \times 16.6 \times 13.9 \cos 106.426°$
   $\Rightarrow c = 24.4799$. Therefore the large triangle ABC has
   $a = 13.9, b = 16.6, c = 24.5, \hat{A} = 33°, \hat{B} = 40.6°, \hat{C} = 106°$.

   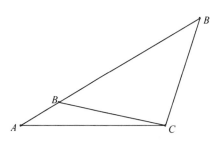

For the small triangle: $\hat{C} = 180° - (33° + 139.426°) = 7.574°$. Then
$c^2 = 16.6^2 + 13.9^2 - 2 \times 16.6 \times 13.9 \cos 7.574° \Rightarrow c = 3.3586$.
So, the small triangle ABC has $a = 13.9$, $b = 16.6$, $c = 3.36$, $\hat{A} = 33°$, $\hat{B} = 139°$, $\hat{C} = 7.57°$.

Exercise 3.10

1. (a) 216 (b) 1.48 (c) 4850

2. $\cos \hat{A} = \dfrac{40.3^2 + 3.86^2 - 2.18^2}{2 \times 4.03 \times 3.86} = 0.8481756 \Rightarrow \hat{A} = 31.986°$

   Area $= \dfrac{1}{2} \times 3.86 \times 4.03 \times \sin 31.986° = 4.12 \text{cm}^2$

3. $\dfrac{15}{\sin \hat{C}} = \dfrac{9}{\sin 30°} \Rightarrow \sin \hat{C} = \dfrac{15 \sin 30°}{9} \Rightarrow \hat{C} = 56.443°, 123.557°$. Therefore $\hat{B} = 93.557°, 26.443°$.

   Therefore, the area of the smaller triangle is $\dfrac{1}{2} \times 15 \times 9 \times \sin 26.443° = 30.1$ and the area of the larger triangle is $\dfrac{1}{2} \times 15 \times 9 \times \sin 93.557° = 67.4$.

4. $\hat{R} = 36°$, using sine rule $\dfrac{8}{\sin 36°} = \dfrac{p}{\sin 75°} \Rightarrow p = 13.1466$.

   Therefore the area is $\dfrac{1}{2} \times 8 \times 13.1466 \times \sin 69° = 49.1 \text{m}^2$.

5. $10\sqrt{3} = \dfrac{1}{2} \times 5 \times 8 \times \sin A \Rightarrow \sin A = \dfrac{\sqrt{3}}{2} \Rightarrow A = 60°, 120°$. Therefore there are two possible triangles which have the given data. For the triangle with $\hat{A} = 60°$, using cosine rule, $BC^2 = 5^2 + 8^2 - 2 \times 5 \times 8 \times \cos 60° \Rightarrow BC = 7$. For the triangle with $\hat{A} = 120°$, $BC^2 = 5^2 + 8^2 - 2 \times 5 \times 8 \times \cos 120° \Rightarrow BC = \sqrt{129}$. Therefore, the possible values of BC are 7, $\sqrt{129}$.

6. (a) Area $\triangle A_1 B_1 C_1 = \dfrac{1}{2} \times 1 \times 1 \times \sin 60° = \dfrac{\sqrt{3}}{4}$

   (b) Area $\triangle A_2 B_2 C_2$ is equilateral and $A_2 B_2 = B_2 C_2 = C_2 A_2 = \dfrac{1}{2}$ so area
   $\triangle A_2 B_2 C_2 = \dfrac{1}{2} \times \left(\dfrac{1}{2} \times \dfrac{1}{2}\right) \sin 60° = \dfrac{\sqrt{3}}{16}$.

   (c) Area $\triangle A_3 B_3 C_3 = \dfrac{1}{2} \times \left(\dfrac{1}{4} \times \dfrac{1}{4}\right) \sin 60° = \dfrac{\sqrt{3}}{64}$

(d) The areas form a geometric sequence in which $a = \dfrac{\sqrt{3}}{4}$, $r = \dfrac{1}{4}$. Therefore the area of

$$\Delta A_{12}B_{12}C_{12} = u_{12} = ar^{11} = \dfrac{\sqrt{3}}{4}\left(\dfrac{1}{4}\right)^{11} = \dfrac{\sqrt{3}}{4^{12}} = \dfrac{\sqrt{3}}{2^{24}} \Rightarrow n = 24.$$

7. $XY^2 = 16^2 + 20^2 - 2\times 16\times 20\cos 125° = 1023.0889 \Rightarrow XY = 31.9858$
$YZ^2 = 20^2 + 9^2 = 481 \Rightarrow YZ = 21.9317$, $ZX^2 = 16^2 + 9^2 = 337 \Rightarrow ZX = 18.3576$
Therefore, $XY = 32.0$cm, $YZ = 21.9$cm, $ZX = 18.4$cm.

The area of $\Delta XYZ = \dfrac{1}{2}\times ZY \times ZX \times \cos X\hat{Z}Y$.

$\cos X\hat{Z}Y = \dfrac{481 + 337 - 1023.0889}{2\times 21.9317\times 18.3576} = -0.254697 \Rightarrow X\hat{Z}Y = 104.7556°$. Therefore area of

$\Delta XYZ = \dfrac{1}{2}\times 21.9317\times 18.3576\times \sin 104.7556° = 194.668 = 195$cm^2, correct to three significant figures.

Solutions to Unit 4 Exercises

Exercise 4.1

1. (a) 3×1 (b) 3×3 (c) 2×5 (d) 2×2 (e) 1×1 (f) 1×4

2. (a) $\begin{pmatrix} -4 & -1 \\ 2 & 1 \end{pmatrix}$ (b) $\begin{pmatrix} 2 & 4 \\ -1 & -2 \end{pmatrix}$ (c) $\begin{pmatrix} -1 & 1 \\ 1 & 2 \end{pmatrix}$ (d) $\begin{pmatrix} 5 & -2 \\ -1 & 7 \end{pmatrix}$ (e) $\begin{pmatrix} -9 & 7 \\ 4 & -4 \end{pmatrix}$

3. (i) (a) $\mathbf{AB} = \begin{pmatrix} 0 & 2 & 6 \\ -13 & -1 & 10 \end{pmatrix}$ (b) $\mathbf{AC} = \begin{pmatrix} 11 \\ 14 \end{pmatrix}$ (c) Incompatible

   (d) $\mathbf{DB} = \begin{pmatrix} 11 & 5 & 4 \end{pmatrix}$ (e) Incompatible (f) $\mathbf{BE} = \begin{pmatrix} 6 & 17 & -2 \\ -2 & -8 & -6 \end{pmatrix}$

   (g) $\mathbf{BF} = \begin{pmatrix} 9 \\ -2 \end{pmatrix}$ (h) $\mathbf{DA} = \begin{pmatrix} 8 & 25 \end{pmatrix}$ (i) $\mathbf{CD} = \begin{pmatrix} 5 & 2 \\ 15 & 6 \end{pmatrix}$ (j) $\mathbf{DC} = (11)$

   (ii) (a) $\mathbf{ACD} = \begin{pmatrix} 55 & 22 \\ 70 & 28 \end{pmatrix}$ (b) $\mathbf{A}^2 = \begin{pmatrix} 1 & 21 \\ -7 & 22 \end{pmatrix}$ (c) $\mathbf{E}^2 = \begin{pmatrix} 19 & -4 & 32 \\ 0 & 27 & -30 \\ 6 & -12 & 33 \end{pmatrix}$

   (d) $\mathbf{CDA} = \begin{pmatrix} 8 & 25 \\ 24 & 75 \end{pmatrix}$

4. $\mathbf{Q+R} = \begin{pmatrix} 2 \\ 1 \end{pmatrix} + \begin{pmatrix} -3 \\ 2 \end{pmatrix} = \begin{pmatrix} -1 \\ 3 \end{pmatrix}$, $\mathbf{P(Q+R)} = \begin{pmatrix} 1 & -5 \\ 2 & 3 \end{pmatrix}\begin{pmatrix} -1 \\ 3 \end{pmatrix} = \begin{pmatrix} -16 \\ 7 \end{pmatrix}$, $\mathbf{PQ} = \begin{pmatrix} 1 & -5 \\ 2 & 3 \end{pmatrix}\begin{pmatrix} 2 \\ 1 \end{pmatrix} = \begin{pmatrix} -3 \\ 7 \end{pmatrix}$,

   $\mathbf{PR} = \begin{pmatrix} 1 & -5 \\ 2 & 3 \end{pmatrix}\begin{pmatrix} -3 \\ 2 \end{pmatrix} = \begin{pmatrix} -13 \\ 0 \end{pmatrix} \Rightarrow \mathbf{PQ+PR} = \begin{pmatrix} -3 \\ 7 \end{pmatrix} + \begin{pmatrix} -13 \\ 0 \end{pmatrix} = \begin{pmatrix} -16 \\ 7 \end{pmatrix}$.

   Therefore, $\mathbf{P(Q+R) = PQ+PR}$.

5. (a) There are many matrices that satisfy the required condition. For example,
   $\mathbf{P} = \begin{pmatrix} 5 & 0 \\ 1 & 3 \end{pmatrix}$, $\mathbf{Q} = \begin{pmatrix} -2 & 4 \\ 1 & 1 \end{pmatrix} \Rightarrow \mathbf{PQ} = \begin{pmatrix} -10 & 20 \\ 1 & 7 \end{pmatrix}$ and $\mathbf{QP} = \begin{pmatrix} -6 & 12 \\ 6 & 3 \end{pmatrix}$

   (b) Let $\mathbf{B} = \begin{pmatrix} b_1 & b_2 \\ b_3 & b_4 \end{pmatrix}$, then $\mathbf{AB} = \begin{pmatrix} a & 0 \\ 0 & a \end{pmatrix}\begin{pmatrix} b_1 & b_2 \\ b_3 & b_4 \end{pmatrix} = \begin{pmatrix} ab_1 & ab_2 \\ ab_3 & ab_4 \end{pmatrix}$ and

   $\mathbf{BA} = \begin{pmatrix} b_1 & b_2 \\ b_3 & b_4 \end{pmatrix}\begin{pmatrix} a & 0 \\ 0 & a \end{pmatrix} = \begin{pmatrix} b_1 a & b_2 a \\ b_3 a & b_4 a \end{pmatrix} = \begin{pmatrix} ab_1 & ab_2 \\ ab_3 & ab_4 \end{pmatrix} = \mathbf{AB}$.

6. (a) $\mathbf{B} = \begin{pmatrix} -4 & -3 \\ 5 & -6 \end{pmatrix}$ (b) $\begin{pmatrix} 1 & 2 \\ 0 & 1 \end{pmatrix} \begin{pmatrix} b_1 & b_2 \\ b_3 & b_4 \end{pmatrix} = \begin{pmatrix} b_1 + 2b_3 & b_2 + 2b_4 \\ b_3 & b_4 \end{pmatrix} = \begin{pmatrix} 1 & 0 \\ 0 & 1 \end{pmatrix}$, $b_3 = 0$

Equating coefficients, $b_1 + 2b_3 = 1$, $b_2 + 2b_4 = 0$, $b_3 = 0$, $b_4 = 1$

$\Rightarrow b_1 = 1$, $b_3 = 0$, $b_4 = 1$, $b_2 = -2b_4 \Rightarrow b_2 = -2$ and therefore $B = \begin{pmatrix} 1 & -2 \\ 0 & 1 \end{pmatrix}$.

Exercise 4.2

1. (a) 1 (b) −7 (c) 6 (d) 1 (e) 0 (f) 1 (g) $-2a^2$

2. (a) 1 (b) −7 (c) −5 (d) 5 (e) −1.14 (f) 0

3. (a) $\begin{pmatrix} 3 & -2 \\ -7 & 5 \end{pmatrix}$ (b) $\begin{pmatrix} -\frac{5}{7} & \frac{6}{7} \\ \frac{2}{7} & -\frac{1}{7} \end{pmatrix}$ (c) $\begin{pmatrix} \frac{1}{6} & \frac{1}{2} \\ -\frac{1}{6} & \frac{1}{2} \end{pmatrix}$ (d) $\begin{pmatrix} 3 & 2 \\ -2 & -1 \end{pmatrix}$

   (e) No inverse (f) $\begin{pmatrix} \sin\theta & \cos\theta \\ -\cos\theta & \sin\theta \end{pmatrix}$ (g) $\begin{pmatrix} \frac{1}{2a} & \frac{1}{2} \\ \frac{1}{2a^2} & -\frac{1}{2a} \end{pmatrix}$

4. (a) $\begin{pmatrix} 0 & -1 & 1 \\ 1 & 2 & -1 \\ -1 & -1 & 1 \end{pmatrix}$ (b) $\begin{pmatrix} 1 & -\frac{4}{7} & \frac{5}{7} \\ 1 & -\frac{1}{7} & \frac{3}{7} \\ 2 & -1 & 1 \end{pmatrix}$ (c) $\begin{pmatrix} \frac{1}{5} & -\frac{1}{5} & -\frac{1}{5} \\ \frac{3}{5} & -\frac{2}{5} & \frac{8}{5} \\ \frac{1}{5} & \frac{4}{5} & -\frac{1}{5} \end{pmatrix}$ (d) $\begin{pmatrix} \frac{2}{5} & \frac{2}{5} & \frac{1}{5} \\ \frac{2}{5} & \frac{7}{5} & \frac{1}{5} \\ -\frac{3}{5} & \frac{8}{5} & \frac{1}{5} \end{pmatrix}$

   (e) $\begin{pmatrix} -0.50 & -0.62 & 0.77 \\ -1.96 & -0.02 & 1.47 \\ 0.82 & 0.16 & 0.02 \end{pmatrix}$ (f) No inverse

5. (a) $\det \mathbf{A} = 8 \Rightarrow \mathbf{A}^{-1} = \frac{1}{8} \begin{pmatrix} 2 & 2 \\ -1 & 3 \end{pmatrix}$.

   $\mathbf{A}^{-1}(\mathbf{AB}) = \mathbf{A}^{-1}(8\mathbf{I}) \Rightarrow (\mathbf{A}^{-1}\mathbf{A})\mathbf{B} = 8\mathbf{A}^{-1} \Rightarrow \mathbf{B} = 8\mathbf{A}^{-1} = \begin{pmatrix} 2 & 2 \\ -1 & 3 \end{pmatrix}$

   (b) $\mathbf{B}^{-1} = \begin{pmatrix} -\frac{4}{7} & -1 & \frac{6}{7} \\ \frac{5}{7} & 1 & -\frac{4}{7} \\ \frac{2}{7} & 1 & -\frac{3}{7} \end{pmatrix}$. $(\mathbf{AB})\mathbf{B}^{-1} = (28\mathbf{I})\mathbf{B}^{-1} \Rightarrow \mathbf{A}(\mathbf{BB}^{-1}) = 28\mathbf{B}^{-1} \Rightarrow \mathbf{AI} = 28\mathbf{B}^{-1} \Rightarrow \mathbf{A} = 28\mathbf{B}^{-1}$

$$\Rightarrow \mathbf{A} = 28\begin{pmatrix} -\frac{4}{7} & -1 & \frac{6}{7} \\ \frac{5}{7} & 1 & -\frac{4}{7} \\ \frac{2}{7} & 1 & -\frac{3}{7} \end{pmatrix} = \begin{pmatrix} -16 & -28 & 24 \\ 20 & 28 & -16 \\ 8 & 28 & -12 \end{pmatrix}$$

6. (i) (a) $\mathbf{AB} = \begin{pmatrix} -5 & -7 \\ -1 & 1 \end{pmatrix}$ (b) $(\mathbf{AB})^{-1} = \begin{pmatrix} -\frac{1}{12} & -\frac{7}{12} \\ -\frac{1}{12} & \frac{5}{12} \end{pmatrix}$ (c) $\mathbf{A}^{-1} = \begin{pmatrix} -\frac{1}{2} & \frac{5}{2} \\ -1 & 4 \end{pmatrix}$ (d) $\mathbf{B}^{-1} = \begin{pmatrix} -\frac{11}{6} & 1 \\ \frac{1}{6} & 0 \end{pmatrix}$

(e) $\mathbf{B}^{-1}\mathbf{A}^{-1} = -\frac{1}{6}\begin{pmatrix} 11 & -6 \\ -1 & 0 \end{pmatrix}\frac{1}{2}\begin{pmatrix} -1 & 5 \\ -2 & 8 \end{pmatrix} = \begin{pmatrix} -\frac{1}{12} & -\frac{7}{12} \\ -\frac{1}{12} & \frac{5}{12} \end{pmatrix}$. Therefore, $(\mathbf{AB})^{-1} = \mathbf{B}^{-1}\mathbf{A}^{-1}$.

(ii) $\mathbf{PQ} = \begin{pmatrix} -5 & 4 & -8 \\ -2 & 3 & -9 \\ 2 & 0 & -2 \end{pmatrix} \Rightarrow (\mathbf{PQ})^{-1} = \begin{pmatrix} 0.6 & -0.8 & 1.2 \\ 2.2 & -2.6 & 2.9 \\ 0.6 & -0.8 & 0.7 \end{pmatrix}$

$\mathbf{P}^{-1} = \begin{pmatrix} -0.2 & -0.4 & 0.6 \\ 0.6 & -0.8 & 1.2 \\ 0.8 & -1.4 & 1.6 \end{pmatrix}$, $\mathbf{Q}^{-1} = \begin{pmatrix} 0 & 1 & 0 \\ -1.5 & 0.5 & 2 \\ -0.5 & -0.5 & 1 \end{pmatrix}$

$\mathbf{Q}^{-1}\mathbf{P}^{-1} = \begin{pmatrix} 0 & 1 & 0 \\ -1.5 & 0.5 & 2 \\ -0.5 & -0.5 & 1 \end{pmatrix}\begin{pmatrix} -0.2 & -0.4 & 0.6 \\ 0.6 & -0.8 & 1.2 \\ 0.8 & -1.4 & 1.6 \end{pmatrix} = \begin{pmatrix} 0.6 & -0.8 & 1.2 \\ 2.2 & -2.6 & 2.9 \\ 0.6 & -0.8 & 0.7 \end{pmatrix} = (\mathbf{PQ})^{-1}$

7. (a) $\begin{pmatrix} 1 & 1 \\ -2 & -1 \end{pmatrix}$ (b) $\begin{pmatrix} 0 & -1 \\ 1 & 0 \end{pmatrix}$ (c) $\begin{pmatrix} 0 & -1 \\ 5 & 4 \end{pmatrix}$

8. (a) $\begin{pmatrix} 0 & \frac{1}{2} \\ 1 & 0 \end{pmatrix}$ (b) $\begin{pmatrix} 2 & 0 \\ -1 & 0 \end{pmatrix}$ (c) $\begin{pmatrix} 4 & 2 \\ 5 & 4 \end{pmatrix}$

Exercise 4.3

1. (a) $\mathbf{M} = \begin{pmatrix} 5 & 3 \\ -1 & 1 \end{pmatrix}$, $\det \mathbf{M} = 8 \Rightarrow \mathbf{M}^{-1} = \frac{1}{8}\begin{pmatrix} 1 & -3 \\ 1 & 5 \end{pmatrix} \Rightarrow \begin{pmatrix} x \\ y \end{pmatrix} = \frac{1}{8}\begin{pmatrix} 1 & -3 \\ 1 & 5 \end{pmatrix}\begin{pmatrix} 19 \\ 1 \end{pmatrix}$

$= \frac{1}{8}\begin{pmatrix} 16 \\ 24 \end{pmatrix} = \begin{pmatrix} 2 \\ 3 \end{pmatrix} \Rightarrow x = 2, y = 3$

(b) $\mathbf{M} = \begin{pmatrix} 2 & 3 \\ 4 & 7 \end{pmatrix}$, $\det \mathbf{M} = 2 \Rightarrow \mathbf{M}^{-1} = \frac{1}{2}\begin{pmatrix} 7 & -3 \\ -4 & 2 \end{pmatrix} \Rightarrow \begin{pmatrix} x \\ y \end{pmatrix} = \frac{1}{2}\begin{pmatrix} 7 & -3 \\ -4 & 2 \end{pmatrix}\begin{pmatrix} 17 \\ 45 \end{pmatrix}$

$$= \frac{1}{2}\begin{pmatrix} -16 \\ 22 \end{pmatrix} = \begin{pmatrix} -8 \\ 11 \end{pmatrix} \Rightarrow x = -8, \; y = 11$$

(c) $\mathbf{M} = \begin{pmatrix} 8 & -4 \\ 7 & -3 \end{pmatrix}$, $\det \mathbf{M} = 4 \Rightarrow \mathbf{M}^{-1} = \frac{1}{4}\begin{pmatrix} -3 & 4 \\ -7 & 8 \end{pmatrix} \Rightarrow \begin{pmatrix} x \\ y \end{pmatrix} = \frac{1}{4}\begin{pmatrix} -3 & 4 \\ -7 & 8 \end{pmatrix}\begin{pmatrix} -20 \\ 53 \end{pmatrix}$

$$= \frac{1}{4}\begin{pmatrix} 272 \\ 564 \end{pmatrix} = \begin{pmatrix} 68 \\ 141 \end{pmatrix} \Rightarrow x = 68, \; y = 141$$

(d) $\mathbf{M} = \begin{pmatrix} -8 & -7 \\ 10 & 9 \end{pmatrix}$, $\det \mathbf{M} = -2 \Rightarrow \mathbf{M}^{-1} = -\frac{1}{2}\begin{pmatrix} 9 & 7 \\ -10 & -8 \end{pmatrix} \Rightarrow \begin{pmatrix} x \\ y \end{pmatrix} = -\frac{1}{2}\begin{pmatrix} 9 & 7 \\ -10 & -8 \end{pmatrix}\begin{pmatrix} 1 \\ -3 \end{pmatrix}$

$$= -\frac{1}{2}\begin{pmatrix} -12 \\ 14 \end{pmatrix} = \begin{pmatrix} -6 \\ 7 \end{pmatrix} \Rightarrow x = 6, \; y = -7$$

(e) $\mathbf{M} = \begin{pmatrix} -5 & 2 \\ 3 & -8 \end{pmatrix}$ then $\det \mathbf{M} = 34 \Rightarrow \mathbf{M}^{-1} = \frac{1}{34}\begin{pmatrix} -8 & -2 \\ -3 & -5 \end{pmatrix} \Rightarrow \begin{pmatrix} x \\ y \end{pmatrix} = -\frac{1}{34}\begin{pmatrix} 8 & 2 \\ 3 & 5 \end{pmatrix}\begin{pmatrix} 9 \\ 2 \end{pmatrix}$

$$= -\frac{1}{34}\begin{pmatrix} 76 \\ 37 \end{pmatrix} = \begin{pmatrix} -2.235 \\ -1.088 \end{pmatrix}$$

(f) $\mathbf{M} = \begin{pmatrix} 19 & 31 \\ -11 & 14 \end{pmatrix}$ then $\det \mathbf{M} = 607 \Rightarrow \mathbf{M}^{-1} = \frac{1}{607}\begin{pmatrix} 14 & -31 \\ 11 & 19 \end{pmatrix} \Rightarrow \begin{pmatrix} x \\ y \end{pmatrix} = \frac{1}{607}\begin{pmatrix} 14 & -31 \\ 11 & 19 \end{pmatrix}\begin{pmatrix} 77 \\ 39 \end{pmatrix}$

$$= \frac{1}{607}\begin{pmatrix} -131 \\ 1588 \end{pmatrix} = \begin{pmatrix} -0.216 \\ 2.616 \end{pmatrix}$$

2. (a) $\begin{pmatrix} 11 & 9 \\ 5 & 4 \end{pmatrix}\begin{pmatrix} x \\ y \end{pmatrix} = \begin{pmatrix} 12 \\ 5 \end{pmatrix}$, $\mathbf{M} = \begin{pmatrix} 11 & 9 \\ 5 & 4 \end{pmatrix} \Rightarrow \mathbf{M}^{-1} = \begin{pmatrix} -4 & 9 \\ 5 & -11 \end{pmatrix} \Rightarrow \begin{pmatrix} x \\ y \end{pmatrix} = \begin{pmatrix} -4 & 9 \\ 5 & -11 \end{pmatrix}\begin{pmatrix} 12 \\ 5 \end{pmatrix} = \begin{pmatrix} -3 \\ 5 \end{pmatrix}$

$\Rightarrow x = -3, \; y = 5$

(b) $\begin{pmatrix} 7 & 9 \\ 8 & -7 \end{pmatrix}\begin{pmatrix} x \\ y \end{pmatrix} = \begin{pmatrix} 29 \\ 85 \end{pmatrix}$, $\mathbf{M} = \begin{pmatrix} 7 & 9 \\ 8 & -7 \end{pmatrix} \Rightarrow \mathbf{M}^{-1} = \frac{1}{-121}\begin{pmatrix} -7 & -9 \\ -8 & 7 \end{pmatrix} = \frac{1}{121}\begin{pmatrix} 7 & 9 \\ 8 & -7 \end{pmatrix}$

$\Rightarrow \begin{pmatrix} x \\ y \end{pmatrix} = \frac{1}{121}\begin{pmatrix} 7 & 9 \\ 8 & -7 \end{pmatrix}\begin{pmatrix} 29 \\ 85 \end{pmatrix} = \frac{1}{121}\begin{pmatrix} 968 \\ -363 \end{pmatrix} = \begin{pmatrix} 8 \\ -3 \end{pmatrix} \Rightarrow x = 8, \; y = -3$

(c) $\begin{pmatrix} \frac{3}{4} & -\frac{2}{3} \\ \frac{2}{3} & \frac{1}{3} \end{pmatrix}\begin{pmatrix} x \\ y \end{pmatrix} = \begin{pmatrix} \frac{1}{12} \\ \frac{13}{60} \end{pmatrix}$ $\mathbf{M} = \begin{pmatrix} \frac{3}{4} & -\frac{2}{3} \\ \frac{2}{3} & \frac{1}{3} \end{pmatrix} \Rightarrow \mathbf{M}^{-1} = \frac{36}{25}\begin{pmatrix} \frac{1}{3} & \frac{2}{3} \\ -\frac{2}{3} & \frac{3}{4} \end{pmatrix} \Rightarrow \begin{pmatrix} x \\ y \end{pmatrix} = \frac{36}{25}\begin{pmatrix} \frac{1}{3} & \frac{2}{3} \\ -\frac{2}{3} & \frac{3}{4} \end{pmatrix}\begin{pmatrix} \frac{1}{12} \\ \frac{13}{60} \end{pmatrix} = \frac{36}{25}\begin{pmatrix} \frac{31}{180} \\ \frac{77}{720} \end{pmatrix} = \begin{pmatrix} \frac{31}{125} \\ \frac{77}{500} \end{pmatrix}$

$\Rightarrow x = \frac{31}{125}, \; y = \frac{77}{500}$

(d) $\begin{pmatrix} 7 & -13 \\ 5 & -1 \end{pmatrix}\begin{pmatrix} x \\ y \end{pmatrix} = \begin{pmatrix} -11 \\ 17 \end{pmatrix}$ $\mathbf{M} = \begin{pmatrix} 7 & -13 \\ 5 & -1 \end{pmatrix} \Rightarrow \mathbf{M}^{-1} = \frac{1}{58}\begin{pmatrix} -1 & 13 \\ -5 & 7 \end{pmatrix} \Rightarrow \begin{pmatrix} x \\ y \end{pmatrix} = \frac{1}{58}\begin{pmatrix} -1 & 13 \\ -5 & 7 \end{pmatrix}\begin{pmatrix} -11 \\ 17 \end{pmatrix}$

$$= \frac{1}{58}\begin{pmatrix} 232 \\ 174 \end{pmatrix} = \begin{pmatrix} 4 \\ 3 \end{pmatrix} \Rightarrow x = 4, y = 3$$

(e) $\begin{pmatrix} 14 & -22 \\ 10 & -27 \end{pmatrix}\begin{pmatrix} x \\ y \end{pmatrix} = \begin{pmatrix} 61 \\ 44 \end{pmatrix}, \mathbf{M} = \begin{pmatrix} 14 & -22 \\ 10 & -27 \end{pmatrix} \Rightarrow \mathbf{M}^{-1} = \frac{1}{-158}\begin{pmatrix} -27 & 22 \\ -10 & 14 \end{pmatrix}$

$\Rightarrow \begin{pmatrix} x \\ y \end{pmatrix} = \frac{1}{158}\begin{pmatrix} 27 & -22 \\ 10 & -14 \end{pmatrix}\begin{pmatrix} 61 \\ 44 \end{pmatrix} \Rightarrow = \frac{1}{158}\begin{pmatrix} 679 \\ -6 \end{pmatrix} \Rightarrow x = 4.30, y = -0.0380$

(f) $\begin{pmatrix} -8 & 1 \\ 2 & 9 \end{pmatrix}\begin{pmatrix} x \\ y \end{pmatrix} = \begin{pmatrix} 19 \\ 15 \end{pmatrix} \mathbf{M} = \begin{pmatrix} -8 & 1 \\ 2 & 9 \end{pmatrix} \Rightarrow \mathbf{M}^{-1} = -\frac{1}{74}\begin{pmatrix} 9 & -1 \\ -2 & -8 \end{pmatrix} \Rightarrow \begin{pmatrix} x \\ y \end{pmatrix} = \frac{1}{74}\begin{pmatrix} -9 & 1 \\ 2 & 8 \end{pmatrix}\begin{pmatrix} 19 \\ 15 \end{pmatrix}$

$= \frac{1}{74}\begin{pmatrix} -156 \\ 158 \end{pmatrix} \Rightarrow x = -2.11, y = 2.14$

(g) $\begin{pmatrix} 0.33 & -1.41 \\ 1.06 & 0.56 \end{pmatrix}\begin{pmatrix} x \\ y \end{pmatrix} = \begin{pmatrix} 7.39 \\ 2.55 \end{pmatrix} \mathbf{M} = \begin{pmatrix} 0.33 & -1.41 \\ 1.06 & 0.56 \end{pmatrix} \Rightarrow \mathbf{M}^{-1} = \frac{1}{1.68}\begin{pmatrix} 0.56 & 1.41 \\ -1.06 & 0.33 \end{pmatrix}$

$\Rightarrow \begin{pmatrix} x \\ y \end{pmatrix} = \frac{1}{1.68}\begin{pmatrix} 0.56 & 1.41 \\ -1.06 & 0.33 \end{pmatrix}\begin{pmatrix} 7.39 \\ 2.55 \end{pmatrix} = \frac{1}{1.68}\begin{pmatrix} 7.7339 \\ -6.9919 \end{pmatrix} \Rightarrow x = 4.61, y = -4.16$

3. $\mathbf{M}\begin{pmatrix} x \\ y \end{pmatrix} = \begin{pmatrix} u \\ v \end{pmatrix} \Rightarrow \mathbf{M}^{-1}\left(\mathbf{M}\begin{pmatrix} x \\ y \end{pmatrix}\right) = \mathbf{M}^{-1}\begin{pmatrix} u \\ v \end{pmatrix}$ by multiplying each side by $\mathbf{M}^{-1}$ which exists,

because $\det \mathbf{M} \neq 0$. $(\mathbf{M}^{-1}\mathbf{M})\begin{pmatrix} x \\ y \end{pmatrix} = \mathbf{M}^{-1}\begin{pmatrix} u \\ v \end{pmatrix}$ by using the associative rule on the left hand side.

But $\mathbf{M}^{-1}\mathbf{M} = \mathbf{I}$ and $\mathbf{I}\begin{pmatrix} x \\ y \end{pmatrix} = \begin{pmatrix} x \\ y \end{pmatrix}$. Therefore $\begin{pmatrix} x \\ y \end{pmatrix} = \mathbf{M}^{-1}\begin{pmatrix} u \\ v \end{pmatrix}$. If $\mathbf{M} = \begin{pmatrix} a & b \\ c & d \end{pmatrix}$ then

$\det \mathbf{M} = ad - bc = \Delta$ and $\mathbf{M}^{-1} = \frac{1}{\Delta}\begin{pmatrix} d & -b \\ -c & a \end{pmatrix} \begin{pmatrix} x \\ y \end{pmatrix} = \frac{1}{\Delta}\begin{pmatrix} d & -b \\ -c & a \end{pmatrix}\begin{pmatrix} u \\ v \end{pmatrix}$

$\Rightarrow x = \frac{ud - bv}{\Delta}, y = \frac{-cu + av}{\Delta}.$

4. (a) $x = 3, y = -2, z = 4$ (b) $x = 2, y = 1, z = -1$

(c) $x = 3, y = 7, z = 11$ (d) $x = 1.16, y = 6.96, z = -2.04$

(e) $x = 1.77, y = 0.391, z = 1.31$ (f) $x = -4.89, y = 7.15, z = 2.49$

5. $\mathbf{A}^{-1} = \begin{pmatrix} 1 & -1 & 3 \\ -0.4 & 0.7 & -1.7 \\ -0.6 & 0.8 & -1.8 \end{pmatrix} \Rightarrow \begin{pmatrix} x \\ y \\ z \end{pmatrix} = \mathbf{A}^{-1}\begin{pmatrix} -35 \\ 53 \\ 33 \end{pmatrix} = \begin{pmatrix} 11 \\ -5 \\ 4 \end{pmatrix} \Rightarrow x = 11, y = -5, z = 4$

6. (a) $\begin{pmatrix} 4 & 3 & 5 \\ 2 & -4 & -1 \\ 1 & 6 & -5 \end{pmatrix} \begin{pmatrix} x \\ y \\ z \end{pmatrix} = \begin{pmatrix} 42 \\ 1 \\ 20 \end{pmatrix} \Rightarrow \begin{pmatrix} x \\ y \\ z \end{pmatrix} = \begin{pmatrix} 7 \\ 3 \\ 1 \end{pmatrix} \Rightarrow x = 7,\ y = 3,\ z = 1$

(b) $\begin{pmatrix} 1 & 1 & -3 \\ 2 & -6 & 1 \\ 3 & -5 & -7 \end{pmatrix} \begin{pmatrix} x \\ y \\ z \end{pmatrix} = \begin{pmatrix} -14 \\ 25 \\ -4 \end{pmatrix} \Rightarrow \begin{pmatrix} x \\ y \\ z \end{pmatrix} = \begin{pmatrix} -1 \\ -4 \\ 3 \end{pmatrix} \Rightarrow x = -1,\ y = -4,\ z = 3$

(c) $\begin{pmatrix} 31 & -15 & 9 \\ 16 & 10 & -11 \\ 23 & 4 & 18 \end{pmatrix} \begin{pmatrix} x \\ y \\ z \end{pmatrix} = \begin{pmatrix} 174 \\ -51 \\ 231 \end{pmatrix} \Rightarrow \begin{pmatrix} x \\ y \\ z \end{pmatrix} = \begin{pmatrix} 3 \\ 0 \\ 9 \end{pmatrix} \Rightarrow x = 3,\ y = 0,\ z = 9$

7. (a) $\begin{pmatrix} 5 & 3 & 11 \\ 9 & -2 & -1 \\ 17 & 19 & 11 \end{pmatrix} \begin{pmatrix} x \\ y \\ z \end{pmatrix} = \begin{pmatrix} 38 \\ 15 \\ 47 \end{pmatrix} \Rightarrow x = 1.81,\ y = -0.793,\ z = 2.85$

(b) $\begin{pmatrix} 1 & -1 & 1 \\ 2 & 2 & -1 \\ 1 & 0 & -4 \end{pmatrix} \begin{pmatrix} x \\ y \\ z \end{pmatrix} = \begin{pmatrix} 15 \\ 11 \\ 19 \end{pmatrix} \Rightarrow x = 10.8,\ y = -6.29,\ z = -2.06$

(c) $\begin{pmatrix} -0.716 & 1.315 & 0.971 \\ 1.134 & 2.016 & -0.116 \\ 0.389 & -0.088 & 1.411 \end{pmatrix} \begin{pmatrix} x \\ y \\ z \end{pmatrix} = \begin{pmatrix} 7.613 \\ 5.303 \\ 8.017 \end{pmatrix} \Rightarrow x = 1.11,\ y = 2.32,\ z = 5.52$

Solutions to Unit 5 Exercises

Exercise 5.1

1. (a) 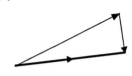 (b)  (c)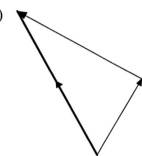

2. (a)  (b) 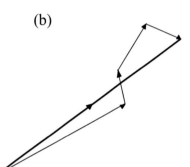 (c)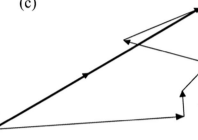

3. $|\vec{v}| = \sqrt{36^2 + 10^2} = 37.4$, $\tan\theta = 5/18 \Rightarrow \theta = 15.5°$

   So the velocity of the harpoon is $37.4\,\mathrm{ms^{-1}}$, at an angle west of north of $15.5°$.

   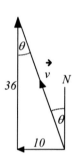

4. (a)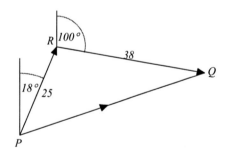

   (b) $P\hat{R}Q = 18° + (180° - 100°) = 98°$.

   Using cosine rule
   $PQ^2 = 25^2 + 38^2 - 2 \times 25 \times 38 \times \cos 98°$
   $\Rightarrow PQ = 48.3056$.

   Using sine rule
   $\dfrac{38}{\sin R\hat{P}Q} = \dfrac{48.3056}{\sin 98°} \Rightarrow \sin R\hat{P}Q = 0.77900$
   and as $R\hat{P}Q < 90°$, $R\hat{P}Q = 51.17°$.

   Therefore $\overrightarrow{PQ}$ has magnitude 48.3km and direction $18° + 51.2° = 69.2°$ east of north.

5. (a)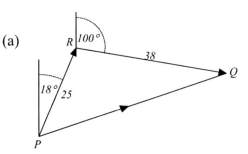

   (b) $|\vec{v}| = \sqrt{0.83^2 - 0.48^2} = 0.677$. Therefore Sheila's speed relative to the bank is 0.677 ms^{-1}.

   (c) As the width of the river is 35 m and Sheila is crossing the river at 0.677 ms^{-1}, she will take $\dfrac{35}{0.677} = 51.70$ seconds, or approximately 52 seconds, to cross.

Exercise 5.2

1. (a) $\overrightarrow{PQ}$ (b) $\overrightarrow{OQ}$ (c) $\overrightarrow{OP} - \overrightarrow{OQ} = \overrightarrow{QP}$

   (d) $\overrightarrow{PR} + \_ + \overrightarrow{OQ} = \overrightarrow{OR} \Rightarrow \_ + \overrightarrow{OQ} = \overrightarrow{OR} - \overrightarrow{PR} = \overrightarrow{OR} + \overrightarrow{RP} = \overrightarrow{OP}$
   $\Rightarrow \overrightarrow{OQ} + \overrightarrow{QP} = \overrightarrow{OP} \Rightarrow \overrightarrow{PR} + \overrightarrow{QP} + \overrightarrow{OQ} = \overrightarrow{OR}$

2. (a) $\overrightarrow{AD}$ (b) $\overrightarrow{OB} + \overrightarrow{BC} = \overrightarrow{OC}$ (c) $\overrightarrow{AB} + \overrightarrow{BA} = \vec{0}$

3. (a) $\overrightarrow{AC} = \vec{b}$ (b) $\vec{a}$ (c) $\overrightarrow{OC} = \vec{a} + \vec{b}$ (d) $\overrightarrow{AB} = \vec{b} - \vec{a}$

4. (a) $\overrightarrow{AB} = \vec{b} - \vec{a}$ (b) $\overrightarrow{BA} = \vec{a} - \vec{b}$ (c) $\overrightarrow{BC} = \overrightarrow{AB} = \vec{b} - \vec{a}$

   (d) $\overrightarrow{AC} = 2\overrightarrow{AB} = 2\vec{b} - 2\vec{a}$ (e) $\overrightarrow{OC} = \vec{a} + (2\vec{b} - 2\vec{a}) = 2\vec{b} - \vec{a}$

5. (a) $\overrightarrow{AC} = \vec{u} + \vec{v}$ (b) $\overrightarrow{CD} = -\vec{u} + \vec{v}$ (c) $\overrightarrow{AD} = \vec{u} + \vec{v} + \overrightarrow{CD} = 2\vec{v}$

   (d) $\overrightarrow{BF} = -\vec{u} + \overrightarrow{AF} = -\vec{u} + \overrightarrow{CD} = \vec{v} - 2\vec{u}$

6. (a)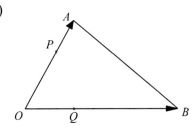

   (b) (i) $\overrightarrow{BA} = \vec{a} - \vec{b}$
       (ii) $\overrightarrow{PB} = \overrightarrow{PO} + \overrightarrow{OB} = -\dfrac{2}{3}\vec{a} + \vec{b} = \vec{b} - \dfrac{2}{3}\vec{a}$
       (iii) $\overrightarrow{AQ} = \overrightarrow{AO} + \overrightarrow{OQ} = -\vec{a} + \dfrac{1}{3}\vec{b}$
       (iv) $\overrightarrow{PQ} = \overrightarrow{PO} + \overrightarrow{OQ} = -\dfrac{2}{3}\vec{a} + \dfrac{1}{3}\vec{b}$

7. (a) $\overrightarrow{AB} = \overrightarrow{AO} + \overrightarrow{OB} = \overrightarrow{OB} - \overrightarrow{OA} = \vec{b} - \vec{a}$

   (b) $\vec{a} + \dfrac{1}{2}\vec{c} = \dfrac{4}{5}\vec{b} \Rightarrow \vec{c} = \dfrac{8}{5}\vec{b} - 2\vec{a}$. Therefore, $\overrightarrow{BC} = \overrightarrow{BO} + \overrightarrow{OC} = \overrightarrow{OC} - \overrightarrow{OB} = \vec{c} - \vec{b}$

$$= \left(\frac{8}{5}\vec{b} - 2\vec{a}\right) - \vec{b} = \frac{3}{5}\vec{b} - 2\vec{a}$$

(c) $\overrightarrow{CA} = \overrightarrow{CO} + \overrightarrow{OA} = -\vec{c} + \vec{a} = \vec{a} - \left(\frac{8}{5}\vec{b} - 2\vec{a}\right) = 3\vec{a} - \frac{8}{5}\vec{b}$

(d) $\overrightarrow{OA} + \overrightarrow{AB} + \overrightarrow{BC} + \overrightarrow{CO} = \vec{a} + (\vec{b} - \vec{a}) + \left(\frac{3}{5}\vec{b} - 2\vec{a}\right) + -\vec{c} = \vec{a} + \vec{b} - \vec{a} + \frac{3}{5}\vec{b} - 2\vec{a} - \left(\frac{8}{5}\vec{b} - 2\vec{a}\right) = \vec{0}$

8. (a) $\overrightarrow{AB} = \vec{b} - \vec{a}$ (b) $\overrightarrow{OP} = \overrightarrow{OA} + \overrightarrow{AP} = \vec{a} + \frac{1}{2}\overrightarrow{AB} = \vec{a} + \frac{1}{2}(\vec{b} - \vec{a}) = \frac{1}{2}(\vec{a} + \vec{b})$

(c) $\overrightarrow{AQ} = \overrightarrow{AO} + \overrightarrow{OQ} = -\vec{a} + \frac{7}{8}\vec{b}$ (d) $\overrightarrow{QR} = \overrightarrow{QO} + \overrightarrow{OR} = -\frac{7}{8}\vec{b} + \frac{1}{2}\vec{a}$

Exercise 5.3

1.

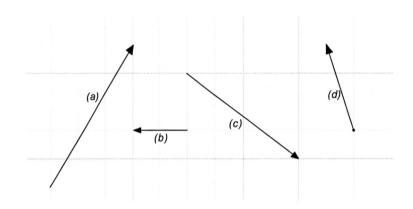

2.

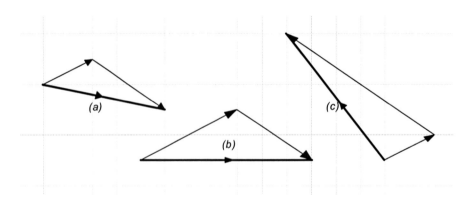

3. (a) (i) $\vec{u} + \vec{v} = \begin{pmatrix} 4 \\ 3 \end{pmatrix} + \begin{pmatrix} -2 \\ 7 \end{pmatrix} = \begin{pmatrix} 2 \\ 10 \end{pmatrix}$

(ii) $2\vec{u}+3\vec{v}=2\begin{pmatrix}4\\3\end{pmatrix}+3\begin{pmatrix}-2\\7\end{pmatrix}=\begin{pmatrix}8\\6\end{pmatrix}+\begin{pmatrix}-6\\21\end{pmatrix}=\begin{pmatrix}2\\27\end{pmatrix}$

(iii) $\dfrac{1}{2}(\vec{u}-\vec{v})=\dfrac{1}{2}\left(\begin{pmatrix}4\\3\end{pmatrix}-\begin{pmatrix}-2\\7\end{pmatrix}\right)=\dfrac{1}{2}\begin{pmatrix}6\\-4\end{pmatrix}=\begin{pmatrix}3\\-2\end{pmatrix}$

(b) (i) $2\vec{u}+\vec{v}=2\left(5\vec{i}-2\vec{j}+\vec{k}\right)+\left(3\vec{i}+\vec{j}+\vec{k}\right)=13\vec{i}-3\vec{j}+3\vec{k}$

(ii) $-\vec{u}+2\vec{v}=-5\vec{i}+2\vec{j}-\vec{k}+2\left(3\vec{i}+\vec{j}+\vec{k}\right)=\vec{i}+4\vec{j}+\vec{k}$

(iii) $\dfrac{1}{3}\vec{u}-\dfrac{1}{2}\vec{v}=\dfrac{1}{3}\left(5\vec{i}-2\vec{j}+\vec{k}\right)-\dfrac{1}{2}\left(3\vec{i}+\vec{j}+\vec{k}\right)=\dfrac{1}{6}\vec{i}-\dfrac{7}{6}\vec{j}-\dfrac{1}{6}\vec{k}$

4. $|\vec{a}|=3|\vec{b}|\Rightarrow\vec{a}=3\vec{b}$ or $\vec{a}=-3\vec{b}$. If $\vec{a}=3\vec{b}$, then $\vec{b}=\dfrac{1}{3}\vec{a}=\dfrac{1}{3}\begin{pmatrix}-2\\0\\3\end{pmatrix}=\begin{pmatrix}-\frac{2}{3}\\0\\1\end{pmatrix}$.

If $\vec{a}=-3\vec{b}$ then $\vec{b}=-\dfrac{1}{3}\begin{pmatrix}2\\0\\-1\end{pmatrix}=\begin{pmatrix}\frac{2}{3}\\0\\-1\end{pmatrix}$.

5. In each case, let $\vec{v}$ be the vector and $\vec{u}$ the corresponding unit vector.

(a) $|\vec{v}|=\sqrt{8^2+15^2}=17\Rightarrow\vec{u}=\dfrac{1}{17}\begin{pmatrix}8\\15\end{pmatrix}=\begin{pmatrix}\frac{8}{17}\\\frac{15}{17}\end{pmatrix}$

(b) $|\vec{v}|=\sqrt{(-7)^2+3^2}=58\Rightarrow\vec{u}=\dfrac{1}{\sqrt{58}}\begin{pmatrix}-7\\3\end{pmatrix}=\begin{pmatrix}-0.919\\0.394\end{pmatrix}$

(c) $\vec{u}=\begin{pmatrix}1\\0\end{pmatrix}$

(d) $|\vec{v}|=\sqrt{6^2+(-5)^2}=\sqrt{61}\Rightarrow\vec{u}=\dfrac{6}{\sqrt{61}}\vec{i}-\dfrac{5}{\sqrt{61}}\vec{j}=0.768\vec{i}-0.640\vec{j}$

(e) $|\vec{v}|=\sqrt{(-0.43)^2+0.37^2}=0.567274\Rightarrow\vec{u}=-0.758\vec{i}+0.652\vec{j}$

6. (a) (i) $\vec{a}+\vec{b} = \begin{pmatrix} 9 \\ 6 \end{pmatrix} \Rightarrow |\vec{a}+\vec{b}| = \sqrt{9^2+6^2} = 10.8$

(ii) $|\vec{a}| = \sqrt{85}, |\vec{b}| = \sqrt{58} \Rightarrow |\vec{a}|+|\vec{b}| = \sqrt{85}+\sqrt{58} = 16.8$

(b) Therefore as 10.8<16.8, $|\vec{a}+\vec{b}| < |\vec{a}|+|\vec{b}|$

(c) Consider the arbitrary vectors $\vec{a}$ and $\vec{b}$. The diagram shows the sides of a vector triangle in which the lengths of the sides are $|\vec{a}|$, $|\vec{b}|$ and $|\vec{a}+\vec{b}|$. It is clear from the diagram that, in order to form a triangle, $|\vec{a}+\vec{b}| < |\vec{a}|+|\vec{b}|$. The equality occurs when $\vec{a}$ and $\vec{b}$ are parallel. Therefore, in general, $|\vec{a}+\vec{b}| \leq |\vec{a}|+|\vec{b}|$.

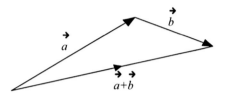

7. $2ab+1=7 \Rightarrow ab=3$; $-3a+4=2 \Rightarrow a=\dfrac{2}{3} \Rightarrow \dfrac{2}{3}b=3 \Rightarrow b=\dfrac{9}{2}$

8. $m(5\vec{i}+\vec{j})+n(3\vec{i}+3\vec{j})=12(-\vec{i}-2\vec{j})$. Therefore, $5m+3n=-12$, $m+3n=-24$. Subtracting we get $4m=12 \Rightarrow m=3$. Then $3+3n=-24 \Rightarrow 1+n=-8 \Rightarrow n=-9$.

9. (a) $\overrightarrow{OA} = \begin{pmatrix} 1 \\ 4 \\ 2 \end{pmatrix}$, $\overrightarrow{OB} = \begin{pmatrix} -3 \\ 1 \\ 6 \end{pmatrix}$, $\overrightarrow{OC} = \begin{pmatrix} 5 \\ 5 \\ -6 \end{pmatrix}$

(b) $|\overrightarrow{OA}| = \sqrt{1^2+4^2+2^2} = \sqrt{21}$; $|\overrightarrow{OB}| = \sqrt{(-3)^2+1^2+6^2} = \sqrt{46}$; $|\overrightarrow{OC}| = \sqrt{86}$

10. (a) $\overrightarrow{AB} = \overrightarrow{OB}-\overrightarrow{OA} = \begin{pmatrix} -3 \\ 6 \end{pmatrix} - \begin{pmatrix} 4 \\ 5 \end{pmatrix} = \begin{pmatrix} -7 \\ 1 \end{pmatrix}$, $\overrightarrow{BC} = \begin{pmatrix} 10 \\ -8 \end{pmatrix}$, $\overrightarrow{CA} = \begin{pmatrix} -3 \\ 7 \end{pmatrix}$

$\overrightarrow{AB}+\overrightarrow{BC}+\overrightarrow{CA} = \begin{pmatrix} -7 \\ 1 \end{pmatrix} + \begin{pmatrix} 10 \\ -8 \end{pmatrix} + \begin{pmatrix} -3 \\ 7 \end{pmatrix} = \begin{pmatrix} 0 \\ 0 \end{pmatrix} = \vec{0}$

(b) $|\overrightarrow{OA}| = \sqrt{41}$, $|\overrightarrow{OB}| = \sqrt{45}$; $\overrightarrow{AB}+2\overrightarrow{BC} = \begin{pmatrix} -7 \\ 1 \end{pmatrix} + 2\begin{pmatrix} 10 \\ -8 \end{pmatrix} = \begin{pmatrix} 13 \\ -15 \end{pmatrix}$

$\Rightarrow |\overrightarrow{AB}+2\overrightarrow{BC}| = \sqrt{13^2+(-15)^2} = 19.8$

Exercise 5.4

1. In each case, assume that the vectors are $\vec{a}$ and $\vec{b}$ and the angle between them is θ.

   (a) $\vec{a} \cdot \vec{b} = 1 \times 2 + 5 \times 2 = 12$, $|\vec{a}| = \sqrt{26}$, $|\vec{b}| = \sqrt{8}$, $\theta = \arccos\left(\dfrac{12}{\sqrt{26}\sqrt{8}}\right) = 33.7°$

   (b) $\vec{a} \cdot \vec{b} = -3 \times 4 + 1 \times -2 = -14$, $|\vec{a}| = \sqrt{10}$, $|\vec{b}| = \sqrt{20}$, $\theta = \arccos\left(\dfrac{-14}{\sqrt{10}\sqrt{20}}\right) = 172°$

   (c) $\vec{a} \cdot \vec{b} = 5 \times 1 + 3 \times -2 = -1$, $|\vec{a}| = \sqrt{34}$, $|\vec{b}| = \sqrt{5}$, $\theta = \arccos\left(\dfrac{-1}{\sqrt{34}\sqrt{5}}\right) = 94.4°$

   (d) $\vec{a} \cdot \vec{b} = 7 \times 0 + 4 \times 3 = 12$, $|\vec{a}| = \sqrt{65}$, $|\vec{b}| = 3$, $\theta = \arccos\left(\dfrac{12}{3\sqrt{65}}\right) = 60.3°$

2. In each case, assume that the vectors are $\vec{a}$ and $\vec{b}$ and the angle between them is θ.

   (a) $\vec{a} \cdot \vec{b} = 1 \times 2 + 1 \times 3 + (-1) \times 2 = 3$; $|\vec{a}| = \sqrt{1^2 + 1^2 + (-1)^2} = \sqrt{3}$, $|\vec{b}| = \sqrt{2^2 + 3^2 + 2^2} = \sqrt{17}$

   $\Rightarrow \theta = \arccos\left(\dfrac{3}{\sqrt{3}\sqrt{17}}\right) = 65.2°$

   (b) $\vec{a} \cdot \vec{b} = 6 \times (-2) + (-2) \times (-5) + 1 \times 4 = 2$, $|\vec{a}| = \sqrt{6^2 + (-2)^2 + 1^2} = \sqrt{41}$,

   $|\vec{b}| = \sqrt{(-2)^2 + (-5)^2 + 4^2} = \sqrt{45} \Rightarrow \theta = \arccos\left(\dfrac{2}{\sqrt{41}\sqrt{45}}\right) = 87.3°$

   (c) $\vec{a} \cdot \vec{b} = (-4) \times 5 + 3 \times 1 + (-3) \times (-6) = 1$, $|\vec{a}| = \sqrt{(-4)^2 + 3^2 + (-3)^2} = \sqrt{34}$,

   $|\vec{b}| = \sqrt{5^2 + 1^2 + (-6)^2} = \sqrt{62} \Rightarrow \theta = \arccos\left(\dfrac{1}{\sqrt{34}\sqrt{62}}\right) = 88.8°$

   (d) $\vec{a} \cdot \vec{b} = 4 \times 3 + 5 \times 4 + 6 \times 6 = 68$, $|\vec{a}| = \sqrt{4^2 + 5^2 + 6^2} = \sqrt{77}$,

   $|\vec{b}| = \sqrt{3^2 + 4^2 + 6^2} = \sqrt{61} \Rightarrow \theta = \arccos\left(\dfrac{68}{\sqrt{77}\sqrt{61}}\right) = 7.16°$

3. (a) $\overrightarrow{AB} = \begin{pmatrix} 0 \\ 2 \end{pmatrix} - \begin{pmatrix} 1 \\ 2 \end{pmatrix} = \begin{pmatrix} -1 \\ 0 \end{pmatrix}$, $\overrightarrow{CD} = \begin{pmatrix} 3 \\ 5 \end{pmatrix} - \begin{pmatrix} -2 \\ 4 \end{pmatrix} = \begin{pmatrix} 5 \\ 1 \end{pmatrix}$ $\overrightarrow{AB} \cdot \overrightarrow{CD} = -5$, $|\overrightarrow{AB}| = 1$, $|\overrightarrow{CD}| = \sqrt{26}$

$$\Rightarrow \theta = \arccos\left(\frac{-5}{1\sqrt{26}}\right) = 169°$$

(b) $\overrightarrow{AC} = \begin{pmatrix} -2 \\ 4 \end{pmatrix} - \begin{pmatrix} 1 \\ 2 \end{pmatrix} = \begin{pmatrix} -3 \\ 2 \end{pmatrix}$; $\overrightarrow{CD} = \begin{pmatrix} 5 \\ 1 \end{pmatrix}$ $\overrightarrow{AC} \cdot \overrightarrow{CD} = -13$, $|\overrightarrow{AC}| = \sqrt{13}$, $|\overrightarrow{CD}| = \sqrt{26}$

$$\Rightarrow \theta = \arccos\left(\frac{-13}{\sqrt{13}\sqrt{26}}\right) = 135°$$

(c) $\overrightarrow{AD} = \begin{pmatrix} 3 \\ 5 \end{pmatrix} - \begin{pmatrix} 1 \\ 2 \end{pmatrix} = \begin{pmatrix} 2 \\ 3 \end{pmatrix}$; $\overrightarrow{BC} = \begin{pmatrix} -2 \\ 4 \end{pmatrix} - \begin{pmatrix} 0 \\ 2 \end{pmatrix} = \begin{pmatrix} -2 \\ 2 \end{pmatrix}$ $\overrightarrow{AD} \cdot \overrightarrow{BC} = 2$, $|\overrightarrow{AD}| = \sqrt{13}$, $|\overrightarrow{BC}| = \sqrt{8}$

$$\Rightarrow \theta = \arccos\left(\frac{2}{\sqrt{8}\sqrt{13}}\right) = 78.7°$$

4. (a) $\overrightarrow{AB} = \begin{pmatrix} 2 \\ 1 \\ 5 \end{pmatrix} - \begin{pmatrix} 0 \\ 1 \\ 3 \end{pmatrix} = \begin{pmatrix} 2 \\ 0 \\ 2 \end{pmatrix}$; $\overrightarrow{DC} = \begin{pmatrix} -3 \\ 2 \\ 0 \end{pmatrix} - \begin{pmatrix} 1 \\ 4 \\ 1 \end{pmatrix} = \begin{pmatrix} -4 \\ -2 \\ -1 \end{pmatrix}$ $\overrightarrow{AB} \cdot \overrightarrow{DC} = -10$, $|\overrightarrow{AB}| = \sqrt{8}$, $|\overrightarrow{DC}| = \sqrt{21}$

$$\Rightarrow \theta = \arccos\left(\frac{-10}{\sqrt{8}\sqrt{21}}\right) = 140°$$

(b) $\overrightarrow{CB} = \begin{pmatrix} 2 \\ 1 \\ 5 \end{pmatrix} - \begin{pmatrix} -3 \\ 2 \\ 0 \end{pmatrix} = \begin{pmatrix} 5 \\ -1 \\ 5 \end{pmatrix}$; $\overrightarrow{AD} = \begin{pmatrix} 1 \\ 4 \\ 1 \end{pmatrix} - \begin{pmatrix} 0 \\ 1 \\ 3 \end{pmatrix} = \begin{pmatrix} 1 \\ 3 \\ -2 \end{pmatrix}$ $\overrightarrow{CB} \cdot \overrightarrow{AD} = -8$, $|\overrightarrow{CB}| = \sqrt{51}$, $|\overrightarrow{AD}| = \sqrt{14}$

$$\Rightarrow \theta = \arccos\left(\frac{-8}{\sqrt{51}\sqrt{14}}\right) = 107°$$

(c) $\overrightarrow{BC} = \begin{pmatrix} -5 \\ 1 \\ -5 \end{pmatrix}$; $\overrightarrow{BD} = \begin{pmatrix} 1 \\ 4 \\ 1 \end{pmatrix} - \begin{pmatrix} 2 \\ 1 \\ 5 \end{pmatrix} = \begin{pmatrix} -1 \\ 3 \\ -4 \end{pmatrix}$ $\overrightarrow{BC} \cdot \overrightarrow{BD} = 28$, $|\overrightarrow{BC}| = \sqrt{51}$, $|\overrightarrow{BD}| = \sqrt{26}$

$$\Rightarrow \theta = \arccos\left(\frac{28}{\sqrt{51}\sqrt{26}}\right) = 39.7°$$

(d) $\overrightarrow{DC} = \begin{pmatrix} -4 \\ -2 \\ -1 \end{pmatrix}$; $\overrightarrow{DA} = \begin{pmatrix} -1 \\ -3 \\ 2 \end{pmatrix}$; $\overrightarrow{DC} \cdot \overrightarrow{DA} = 8$; $|\overrightarrow{DC}| = \sqrt{21}$ $|\overrightarrow{DA}| = \sqrt{14}$

$$\Rightarrow \theta = \arccos\left(\frac{8}{\sqrt{21}\sqrt{14}}\right) = 62.2°$$

5. (a) $\vec{a} \cdot \vec{b} = 0 \Rightarrow 10 \times -8 + 5k = 0 \Rightarrow 5k - 80 = 0 \Rightarrow k = 16$

(b) $\vec{u} \cdot \vec{v} = 0 \Rightarrow 3 \times (-5) + 1 \times (-9) + 8c = 0 \Rightarrow 8c = 24 \Rightarrow c = 3$

6. $\vec{a} \cdot \vec{c} = 3 + 2m$; $\vec{a} \cdot \vec{b} = -5$. Therefore, $\begin{pmatrix} -10 \\ n \end{pmatrix} = (3 + 2m)\begin{pmatrix} 1 \\ -4 \end{pmatrix} + 5\begin{pmatrix} 1 \\ m \end{pmatrix} \Rightarrow -10 = 3 + 2m + 5$

$\Rightarrow 2m = -18 \Rightarrow m = -9$. $n = -4(3 + 2m) + 5m \Rightarrow n = -12 - 8m + 5m = -12 - 3m = -12 + 27 = 15$

So, $m = -9, n = 15$.

7. (a) $\begin{pmatrix} \frac{3}{\sqrt{34}} \\ \frac{-5}{\sqrt{34}} \end{pmatrix}$ or $\begin{pmatrix} \frac{-3}{\sqrt{34}} \\ \frac{5}{\sqrt{34}} \end{pmatrix}$ (b) $\begin{pmatrix} \frac{11}{\sqrt{185}} \\ \frac{8}{\sqrt{185}} \end{pmatrix}$ or $\begin{pmatrix} \frac{-11}{\sqrt{185}} \\ \frac{-8}{\sqrt{185}} \end{pmatrix}$ (c) $\begin{pmatrix} -0.909 \\ 0.418 \end{pmatrix}$ or $\begin{pmatrix} 0.909 \\ -0.418 \end{pmatrix}$

   (d) $\begin{pmatrix} 0 \\ 1 \end{pmatrix}$ or $\begin{pmatrix} 0 \\ -1 \end{pmatrix}$ (e) $-\frac{3}{5}\vec{i} + \frac{4}{5}\vec{j}$ or $\frac{3}{5}\vec{i} - \frac{4}{5}\vec{j}$ (f) $-i$ or i

8. (a)1, (c)2   (b)1, (a)2   (c)1, (b)2   (d)1, (e)2   (e)1, (d)2

9. $\begin{pmatrix} 1 \\ 1 \\ -3 \end{pmatrix} \cdot \begin{pmatrix} 1 \\ -10 \\ -3 \end{pmatrix} = 1 - 10 + 9 = 0$; $\begin{pmatrix} 1 \\ 1 \\ -3 \end{pmatrix} \cdot \begin{pmatrix} 3 \\ 0 \\ 1 \end{pmatrix} = 3 - 3 = 0$; $\begin{pmatrix} 1 \\ -10 \\ -3 \end{pmatrix} \cdot \begin{pmatrix} 3 \\ 0 \\ 1 \end{pmatrix} = 3 - 3 = 0$

10. (a) $\vec{u} \cdot \vec{v} = 21 - 20 - c = 0 \Rightarrow c = 1$ (b) $\vec{u} \cdot \vec{v} = c - 6 + 8 = 0 \Rightarrow c = -2$

    (c) $\vec{u} \cdot \vec{v} = 8 + 5c - 54 = 0 \Rightarrow 5c = 46 \Rightarrow c = 9.2$ (d) $\vec{u} \cdot \vec{v} = 2c - 12 + 2c = 0 \Rightarrow 4c = 12 \Rightarrow c = 3$

11. (a) $\vec{u} \cdot \vec{v} = 3 + 12 + 27 = 42$, $|\vec{u}| = \sqrt{1^2 + 2^2 + (-3)^2} = \sqrt{14}$; $|\vec{v}| = \sqrt{3^2 + 6^2 + (-9)^2} = \sqrt{126}$

    $|\vec{u}||\vec{v}| = \sqrt{14}\sqrt{126} = \sqrt{2}\sqrt{7}\sqrt{2}\sqrt{7}\sqrt{9} = 2 \times 7 \times 3 = 42$. Therefore $\vec{u} \cdot \vec{v} = |\vec{u}||\vec{v}|$.

    (b) $\vec{w} = -\vec{i} - 2\vec{j} + 3\vec{k}$

Exercise 5.5

1. (a) $x = 1 + 2t, y = 2 - t \Rightarrow t = 2 - y \Rightarrow x = 1 + 2(2 - y) \Rightarrow x + 2y - 5 = 0$

   (b) $x = -5 + 4t, y = -2 + 9t \Rightarrow 9x - 4y + 37 = 0$

   (c) $x = 13 + 8(t - 1), y = 5 - 3(t - 1) \Rightarrow 3x + 8y - 79 = 0$

   (d) $x = 3 + 5t, y = -4 + t \Rightarrow x - 5y - 23 = 0$

   (e) $x = -7 + 3t, y = 2 \Rightarrow y = 2$

2. In each case, $\vec{v}_1$ and $\vec{v}_2$ are vectors parallel to the corresponding lines and θ is the angle between the lines.

(a) $\vec{v}_1 = \begin{pmatrix} -1 \\ 1 \end{pmatrix}$, $\vec{v}_2 = \begin{pmatrix} 1 \\ 1 \end{pmatrix} \Rightarrow \vec{v}_1 \cdot \vec{v}_2 = 0 \Rightarrow \theta = 90°$

(b) $\vec{v}_1 = \begin{pmatrix} -3 \\ 2 \end{pmatrix}$, $\vec{v}_2 = \begin{pmatrix} 1 \\ 4 \end{pmatrix} \Rightarrow \vec{v}_1 \cdot \vec{v}_2 = 5 \Rightarrow \cos\theta = \dfrac{5}{\sqrt{17}\sqrt{13}} \Rightarrow \theta = 70.3°$

(c) $\vec{v}_1 = \begin{pmatrix} 6 \\ 5 \end{pmatrix}$, $\vec{v}_2 = \begin{pmatrix} 4 \\ -3 \end{pmatrix} \Rightarrow \vec{v}_1 \cdot \vec{v}_2 = 9 \Rightarrow \cos\theta = \dfrac{9}{5\sqrt{61}} \Rightarrow \theta = 76.7°$

(d) $\vec{v}_1 = \begin{pmatrix} 1 \\ 4 \end{pmatrix}$, $\vec{v}_2 = \begin{pmatrix} -3 \\ 2 \end{pmatrix} \Rightarrow \vec{v}_1 \cdot \vec{v}_2 = 5 \Rightarrow \cos\theta = \dfrac{5}{\sqrt{17}\sqrt{13}} \Rightarrow \theta = 70.3°$

3. (a) $\overrightarrow{AB} = \overrightarrow{OC} = \begin{pmatrix} 6 \\ -3 \end{pmatrix}$, $\overrightarrow{OB} = \overrightarrow{OA} + \overrightarrow{AB} = \begin{pmatrix} -2 \\ 5 \end{pmatrix} + \begin{pmatrix} 6 \\ -3 \end{pmatrix} = \begin{pmatrix} 4 \\ 2 \end{pmatrix}$

(b) $\overrightarrow{AC} = \begin{pmatrix} 6 \\ -3 \end{pmatrix} - \begin{pmatrix} -2 \\ 5 \end{pmatrix} = \begin{pmatrix} 8 \\ -8 \end{pmatrix} = 8\begin{pmatrix} 1 \\ -1 \end{pmatrix}$. Therefore, $k = -1$.

(c) (i) $x = -2 + 1 = -1$, $y = 5 - 1 = 4$, so, after 1 second, P has coordinates $(-1, 4)$.

(ii) $t = 3$

(d) $\begin{pmatrix} x \\ y \end{pmatrix} = \begin{pmatrix} 4 \\ 2 \end{pmatrix} + s\begin{pmatrix} 1 \\ 1 \end{pmatrix}$

(e) Let the point of intersection of L_1 and L_2 be N. At N, $-2 + t = 4 + s$ and $5 - t = 2 + s$. By subtraction, $-7 + 2t = 2 \Rightarrow t = 4.5$. Then, at N $x = -2 + 4.5 = 2.5$ and $y = 5 - 4.5 = 0.5$. Therefore, the coordinates of N are $(2.5, 0.5)$.

Minimum distance, $|\overrightarrow{BN}| = \sqrt{(4-2.5)^2 + (2-0.5)^2} = 2.12$

4. (a) $\vec{r} = (3\vec{i} + 7\vec{j}) + t(6\vec{i} - 6\vec{j})$ (b) When $t = \dfrac{1}{3}$, $\vec{r} = (3\vec{i} + 7\vec{j}) + (-2\vec{i} + 2\vec{j}) = \vec{i} + 9\vec{j}$

(c) $\vec{r} = (\vec{i} + 9\vec{j}) + s(\vec{i} + \vec{j})$ (d) At the x-axis, $y = 0 \Rightarrow 9 + s = 0 \Rightarrow s = -9$. Then $x\vec{i} + y\vec{j} = \vec{i} + s\vec{i} = -8\vec{i} \Rightarrow x = -8$, $y = 0$ and Q is $(-8, 0)$.

5. (a) Antelope's speed is $\sqrt{4^2+3^2}=5\text{ms}^{-1}$.

    (b) Antelope starts when $t=0$ and therefore, starting point is $(-9, 11)$.

    (c) $\vec{r}=\begin{pmatrix}-9\\11\end{pmatrix}+4\begin{pmatrix}4\\3\end{pmatrix}=\begin{pmatrix}7\\23\end{pmatrix}$. Therefore, its position after 4 seconds is $(7, 23)$.

    (d) After 11 seconds, the antelope's position vector is $\begin{pmatrix}x\\y\end{pmatrix}=\begin{pmatrix}-9\\11\end{pmatrix}+11\begin{pmatrix}4\\3\end{pmatrix}=\begin{pmatrix}35\\44\end{pmatrix}$.
    Therefore, the coordinates of the point where the cheetah catches the antelope are $(35, 44)$.

    (e) Vector equation of the path of the cheetah is $\begin{pmatrix}x\\y\end{pmatrix}=\begin{pmatrix}0\\25\end{pmatrix}+t\begin{pmatrix}u\\v\end{pmatrix}$, where $\begin{pmatrix}u\\v\end{pmatrix}$ is the cheetah's velocity. Therefore, if the cheetah is to catch the antelope after 11 seconds $11u=35$ and $-25+11v=44 \Rightarrow u=\frac{35}{11}, v=\frac{69}{11}$. Therefore, the speed of the cheetah is $\sqrt{\left(\frac{35}{11}\right)^2+\left(\frac{69}{11}\right)^2}=7.03\text{ms}^{-1}$.

    (f) Animals do not move in straight lines at constant speed.

6. (a) $\overrightarrow{OP}=t\begin{pmatrix}4\\3\end{pmatrix}$ so $t=1$, $\overrightarrow{OP}=\begin{pmatrix}4\\3\end{pmatrix}$,
    at $t=2$, $\overrightarrow{OP}=\begin{pmatrix}8\\6\end{pmatrix}$ and at $t=5$, $\overrightarrow{OP}=\begin{pmatrix}20\\15\end{pmatrix}$.
    At $t=1$, $\overrightarrow{AP}=\begin{pmatrix}-6\\-10\end{pmatrix}$, at $t=2$, $\overrightarrow{AP}=\begin{pmatrix}-2\\-7\end{pmatrix}$
    and at $t=5$ $\overrightarrow{AP}=\begin{pmatrix}10\\2\end{pmatrix}$.

    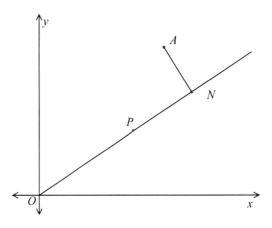

    (b) $\overrightarrow{AP}=\overrightarrow{OP}-\overrightarrow{OA}=t\begin{pmatrix}4\\3\end{pmatrix}-\begin{pmatrix}10\\13\end{pmatrix}=\begin{pmatrix}4t-10\\3t-13\end{pmatrix}$

    (c) $|\overrightarrow{AP}|=15 \Rightarrow \sqrt{(-10+4t)^2+(-13+3t)^2}=15 \Rightarrow t^2-6.32t+1.76=0$
    $\Rightarrow t=0.291969, 6.02803$. Therefore the ship is 15km from the rock for the first time when $t=0.291969$ which is 18 minutes, to the nearest minute, after leaving the harbor. So the first time when the ship is 15 km from rock is 09:18.

(d) The scalar product, $\overrightarrow{AP} \cdot \begin{pmatrix} 4 \\ 3 \end{pmatrix} = \begin{pmatrix} 4t-10 \\ 3t-13 \end{pmatrix} \cdot \begin{pmatrix} 4 \\ 3 \end{pmatrix} = -40 + 16t - 39 + 9t = 25t - 79$.

(e) N is the point closest to the rock (see figure). So $25t - 79 = 0 \Rightarrow t = 3.16$ and the ship is closest to the rock 3 hours 10 minutes after leaving the harbor at 12:10.

Exercise 5.6

1. Assume, in each case, that $\vec{b}_1$ and $\vec{b}_2$ are vectors parallel to the two lines and that θ is the angle between the two vectors.

   (a) $\vec{b}_1 = \vec{i} - 2\vec{j} + 5\vec{k}$, $\vec{b}_2 = 3\vec{i} + 3\vec{j} + 2\vec{k}$, $\vec{b}_1 \cdot \vec{b}_2 = 7$, $|\vec{b}_1| = \sqrt{30}$, $|\vec{b}_2| = \sqrt{22}$

   $\Rightarrow \cos\theta = \dfrac{7}{\sqrt{30}\sqrt{22}} \Rightarrow \theta = 74.2°$

   (b) $\vec{b}_1 = 6\vec{i} + \vec{j} - 3\vec{k}$, $\vec{b}_2 = 5\vec{i} + 4\vec{j} + 3\vec{k}$, $\vec{b}_1 \cdot \vec{b}_2 = 25$, $|\vec{b}_1| = \sqrt{46}$, $|\vec{b}_2| = \sqrt{50}$

   $\Rightarrow \cos\theta = \dfrac{25}{\sqrt{46}\sqrt{50}} \Rightarrow \theta = 58.6°$

   (c) $\vec{b}_1 = -\vec{i} - 8\vec{j} + 2\vec{k}$, $\vec{b}_2 = -3\vec{i} + 8\vec{j} - 7\vec{k}$, $\vec{b}_1 \cdot \vec{b}_2 = -75$, $|\vec{b}_1| = \sqrt{69}$, $|\vec{b}_2| = \sqrt{122}$

   $\Rightarrow \cos\theta = \dfrac{-75}{\sqrt{69}\sqrt{122}} \Rightarrow \theta = 144.8°$. Therefore the acute angle between the lines is $180° - 144.8° = 35.2°$.

2. (a) $\vec{b}_1 = \begin{pmatrix} 3 \\ 5 \\ 1 \end{pmatrix}$, $\vec{b}_2 = \begin{pmatrix} 2 \\ -1 \\ 1 \end{pmatrix} \Rightarrow \vec{b}_1 \cdot \vec{b}_2 = 2$, $|\vec{b}_1| = \sqrt{35}$, $|\vec{b}_2| = \sqrt{6} \Rightarrow \theta = 82.1°$

   (b) $\vec{b}_1 = \begin{pmatrix} -3 \\ -10 \\ 5 \end{pmatrix}$, $\vec{b}_2 = \begin{pmatrix} 0 \\ 1 \\ 0 \end{pmatrix}$, $\Rightarrow \vec{b}_1 \cdot \vec{b}_2 = -10$, $|\vec{b}_1| = \sqrt{134}$, $|\vec{b}_2| = 1 \Rightarrow \theta = 149.8°$

   Therefore the acute angle between the lines is $180° - 149.8° = 30.2°$

   (c) $\vec{b}_1 = \begin{pmatrix} -3 \\ 4 \\ -2 \end{pmatrix}$, $\vec{b}_2 = \begin{pmatrix} 0.7 \\ 1.2 \\ -0.3 \end{pmatrix}$, $\Rightarrow \vec{b}_1 \cdot \vec{b}_2 = 3.3$, $|\vec{b}_1| = \sqrt{29}$, $|\vec{b}_2| = \sqrt{2.02} \Rightarrow \theta = 64.5°$

3. (i) Vector equation of line is $\vec{r} = (3\vec{i} - \vec{j} + \vec{k}) + t(2\vec{i} + 3\vec{j} - 2\vec{k})$. When $x = 7$, $3 + 2t = 7$
$\Rightarrow t = 2$. Therefore, $y = -1 + 3t = 5$ and $z = 1 - 2t = -3$, so that L passes through the point $(7, 5, -3)$.

   (ii) Vector parallel to l is $\begin{pmatrix} 2 \\ 1 \\ 4 \end{pmatrix} - \begin{pmatrix} 1 \\ -3 \\ -2 \end{pmatrix} = \begin{pmatrix} 1 \\ 4 \\ 6 \end{pmatrix}$. Therefore the vector equation of l is

   $\begin{pmatrix} x \\ y \\ z \end{pmatrix} = \begin{pmatrix} 1 \\ -3 \\ -2 \end{pmatrix} + c \begin{pmatrix} 1 \\ 4 \\ 6 \end{pmatrix}$ (there are many other possibilities). When $x = 0$, $1 + c = 0 \Rightarrow c = -1$.

   So $y = -3 + 4(-1) = -7$ and $z = -2 + 6(-1) = -8$ and the point $(0, -7, -8)$ lies on l.

4. (a) $\overrightarrow{OP} = 350\vec{i} - 217\vec{j} + 785\vec{k}$

   (b) $|\vec{v}| = \sqrt{(-9)^2 + 7^2 + (-4)^2} \approx 12.080305$. Therefore the speed of the submarine is 12.1ms^{-1}.

   (c) $\vec{r} = (350\vec{i} - 217\vec{j} + 785\vec{k}) + t(-9\vec{i} + 7\vec{j} - 4\vec{k})$

   (d) At the surface, $-217 + 7t = 0 \Rightarrow t = 31$. Then $x = 350 + 31 \times (-9) = 71$ and $z = 785 + 31 \times (-4) = 661$. Therefore, the submarine surfaces at a point whose coordinates are $(71, 0, 661)$ and the distance from the fishing boat is $\sqrt{(71-10)^2 + (661-600)^2} = 86.3$ meters.

5. (a) Skier's speed is $\sqrt{0.711^2 + (-0.283)^2 + 1.782^2} = 1.94 \text{ms}^{-1}$.

   (b) $\vec{r}(10) = \begin{pmatrix} 5.83 \\ 2.11 \\ 3.74 \end{pmatrix} + 10 \begin{pmatrix} 0.711 \\ -0.283 \\ 1.782 \end{pmatrix} = \begin{pmatrix} 12.94 \\ -0.72 \\ 21.56 \end{pmatrix}$. So, after 10 seconds, the skier's position has coordinates $(12.94, -0.72, 21.56)$.

   (c) $5.83 + 0.711t = 43.51 \Rightarrow t = 53.0$, or, $2.11 - 0.283t = -12.89 \Rightarrow t = 53.0$ or $3.74 + 1.782t = 98.19 \Rightarrow t = 53.0$. Therefore, the time taken is 53.0 seconds.

6. (a) $\overrightarrow{QP} = \overrightarrow{QO} + \overrightarrow{OP} = \overrightarrow{OP} - \overrightarrow{OQ} = (21\vec{i} - 8\vec{j} + 8\vec{k}) - (4\vec{i} + 15\vec{k}) = 17\vec{i} - 8\vec{j} - 7\vec{k}$

   (b) $|\overrightarrow{QP}| = \sqrt{17^2 + (-8)^2 + (-7)^2} \approx 20.05$

$$\overrightarrow{QP} = \frac{1}{20.05}\left(17\vec{i} - 8\vec{j} - 7\vec{k}\right) = 0.848\vec{i} - 0.399\vec{j} - 0.349\vec{k}$$

(c) The time to point P is 20.05 seconds. The time from P to the surface is 8 seconds. Therefore, he needs to hold his breath for 28 seconds.

7. (a) Helen reaches the summit when $2838 - 0.215t = 0 \Rightarrow t = 13200$ or $2178 - 0.165t = 0$ $\Rightarrow t = 13200$. Therefore, the time taken to reach the summit is 13 200 seconds = 220 minutes = 3 hours 40 minutes.

(b) When $t = 13200$ $z = 6150 + 13200 \times 0.131 = 7879.2$. Therefore, the summit has an altitude of 7880 meters.

(c) $7000 - 6150 = 850$. Therefore the time to reach 7000 meters is $\frac{850}{0.1048} = 8110.687$ sec. $= 135.18$ minutes. Russell's speed, using oxygen, is $0.1048 \times 1.3 = 0.13624$ ms^{-1}.
Therefore the time to reach the summit is $\frac{7879.2 - 7000}{0.13624} = 6453.3$ sec. $= 107.56$ minutes.
So, the total time to the summit is $135.18 + 107.56 = 242.74$ minutes = 4 hours 3 minutes.

8. (a) $\overrightarrow{PM} = \overrightarrow{PO} + \overrightarrow{OM} = \overrightarrow{OM} - \overrightarrow{OP}$
$= \left(8\vec{i} + 2\vec{j} + 11\vec{k}\right) - \left[\left(3.5\vec{i} + 8.2\vec{j} + 11.6\vec{k}\right) + t\left(5.3\vec{i} + 0.6\vec{j} + 7.1\vec{k}\right)\right]$
$= \left(4.5\vec{i} - 6.2\vec{j} - 0.6\vec{k}\right) - t\left(5.3\vec{i} + 0.6\vec{j} + 7.1\vec{k}\right)$

(b) $\overrightarrow{PM} \cdot \left(5.3\vec{i} + 0.6\vec{j} + 7.1\vec{k}\right)$
$= \left[\left(4.5\vec{i} - 6.2\vec{j} - 0.6\vec{k}\right) - t\left(5.3\vec{i} + 0.6\vec{j} + 7.1\vec{k}\right)\right] \cdot \left(5.3\vec{i} + 0.6\vec{j} + 7.1\vec{k}\right)$
$= 5.3(4.5 - 5.3t) - 0.6(6.2 + 0.6t) - 7.1(0.6 + 7.1t) = 15.87 - 78.86t$

(c) $15.87 - 78.86t = 0 \Rightarrow t = 0.2012427 \Rightarrow \overrightarrow{PM} = 3.433\vec{i} - 6.321\vec{j} - 2.029\vec{k}$
$\left|\overrightarrow{PM}\right| = \sqrt{\left(3.433^2 + (-6.321)^2 + (-2.029)^2\right)} = 7.47$. So the minimum distance is 7.47 km.

Exercise 5.7

1. (a) $x = 1 - 3p = 7 + q \Rightarrow 3p + q = -6$ .... (1)
$y = 1 + 2p = 5 - q \Rightarrow 2p + q = 4$ .... (2)
Solving (1) and (2) gives $p = -10$, $q = 24$. $z = -4 + p = -4 - 10 = -14$, and $z = 10 - q = 10 - 24 = -14$. Because the values of p and q found in (1) and (2) give a consistent value for z, the lines intersect. The point of intersection is $(31, -19, -14)$.

(b) $x = p = -3 + 2q \Rightarrow p - 2q = -3 \ldots \ldots$ (1)

$y = 2 + p = -1 \ldots \ldots \ldots \ldots \ldots \ldots \ldots$ (2)

Solving (1) and (2) gives $p = -3$ and $q = 0$, $z = 1 + p = 1 - 3 = -2$ and $z = 2 + 3q = 2 + 3 \times 0 = 2 \neq -2$. Therefore the values of p and q found in (1) and (2) give inconsistent values of z, so the lines do not intersect.

(c) $x = 7 + 2p = 2 + 2q \Rightarrow 2p - 2q = -5 \ldots \ldots$ (1)

$y = 2 = 2 + 2q \Rightarrow q = 0 \ldots \ldots \ldots \ldots \ldots$ (2)

Solving (1) and (2) gives $p = -\dfrac{5}{2}$, $q = 0$. $z = 1$, and $z = 1 + q = 1 + 0 = 1$. Because the values of p and q found in (1) and (2) give a consistent value for z, the lines intersect. The point of intersection is $(2, 2, 1)$.

(d) $x = 3p = 6 + 3q \Rightarrow p - q = 2 \ldots \ldots \ldots \ldots$ (1)

$y = -5p = -11 \ldots \ldots \ldots \ldots \ldots \ldots \ldots$ (2)

Solving (1) and (2) gives, $p = \dfrac{11}{5}$, $q = \dfrac{1}{5}$. $z = 8p = \dfrac{88}{5}$ and $z = -q = -\dfrac{1}{5} \neq \dfrac{88}{5}$. The values of p and q found in (1) and (2) give inconsistent values of z, so the lines do not intersect.

2. (a) Equating x components: $1 + pc = -2 + 5q \ldots \ldots$ (1)

Equating y components: $-1 + p = q \ldots \ldots \ldots$ (2)

Equating z components: $0 + p = 1 + 3q \ldots \ldots$ (3)

Solving (2) and (3) $\Rightarrow p - q = 1$, $p - 3q = 1 \Rightarrow q = 0$, $p = 1$. Then, in (1), $1 + c = -2 \Rightarrow c = -3$.

(b) $p = 3 + 2q$, $-p = c - q$, $-1 + p = -3 - 3q$. Therefore, $p = 1$, $q = -1$ and $c = q - p = -2$.

(c) $1 + 2p = 2 + 6q$, $1 + p = cq$, $1 + p = 5 + q$. Therefore $p = \dfrac{23}{4}$, $q = \dfrac{7}{4}$ and $c = \dfrac{27}{7}$.

3. Let the first equation represent l_1 and the second l_2, in each pair.

(a) $(1, 3)$ lies on l_1 so equating the x components of l_1 and l_2 gives $1 = 3 - t \Rightarrow t = 2$ and equating the y components gives $3 = 3 + 2t \Rightarrow t = 0$. Because the values of t are inconsistent, $(1, 3)$ does not lie on l_2 and so the lines are distinct.

(b) $(2, 0, 5)$ lies on l_1. Equating components: $2 = -7 + 3t \Rightarrow t = 3$, $0 = -6 + 2t \Rightarrow t = 3$, $5 = 5$. Because all three equations give consistent results, $(2, 0, 5)$ lies on l_2 so that l_1 and l_2 are coincident.

(c) $(2, -5, -1)$ lies on l_1. Equating components: $2 = -1 + 3t \Rightarrow t = 1$, $-5 = 1 - 6t \Rightarrow t = 1$, $-1 = -13 + 12t \Rightarrow t = 1$. Because all three equations give consistent results, $(2, -5, -1)$ lies on l_2 so that l_1 and l_2 are coincident.

(d) $(0, 4)$ lies on l_1. Equating components: $0 = -1 + 3t \Rightarrow t = \frac{1}{3}$, $4 = -11 + t \Rightarrow t = 15 \neq \frac{1}{3}$
Because the values of t are inconsistent, $(0, 4)$ does not lie on l_2 and so the lines are distinct.

(e) $(3, 10, -1)$ lies on l_1. Equating components: $3 = -5 + 3t \Rightarrow t = \frac{8}{3}$, $10 = 10$,
$-1 = 17 - t \Rightarrow t = 18 \neq \frac{8}{3}$. Because the values of t are inconsistent, $(3, 10, -1)$ does not lie on l_2 and so the lines are distinct.

4. (a) $\vec{r} = \begin{pmatrix} 9.2 \\ 0 \\ 17 \end{pmatrix} + t \begin{pmatrix} 2 \\ 15 \\ 5 \end{pmatrix}$

(b) Let t_1 be the time when the aircraft reaches Q and t_2 be the time when the missile reaches Q. Equating x and y components: $7 + 6t_1 = 9.2 + 2t_2$ ......(1)
$6 = 15t_2$ ....(2)
Solving (1) and (2) gives $6t_1 - 2t_2 = 2.2$, $t_2 = 0.4 \Rightarrow t_1 = 0.5$
Now the z component of the aircraft's path is $15 + 8t_1 = 19$ and the z component of the missile's path is $17 + 5t_2 = 19$, and, because these values are consistent, the paths intersect. The point of intersection Q is $(10, 6, 19)$.

(c) At Q, $t_1 = 0.5$ and $t_2 = 0.4$ so that the aircraft reaches Q after 30 seconds. However, the missile reaches Q after 24 seconds, so that the missile goes in front of the aircraft.

5. (a) $\vec{AB} = (7240\vec{j} + 2260\vec{k}) - (310\vec{i} + 1630\vec{j}) = (-310\vec{i} + 5610\vec{j} + 2260\vec{k})$. This displacement is carried out in 15 minutes = 900 seconds and therefore the velocity of the cable car is
$\frac{1}{900}(-310\vec{i} + 5610\vec{j} + 2260\vec{k}) = -0.344\vec{i} + 6.233\vec{j} + 2.511\vec{k}$ so that $a = -0.344$, $b = 6.23$ and $c = 2.51$.

(b) As the path of the cable car starts at $A(310, 1630, 0)$ and its velocity is
$-0.344\vec{i} + 6.233\vec{j} + 2.511\vec{k}$, the vector equation of the path of the cable car is
$\vec{r} = (310\vec{i} + 1630\vec{j}) + t(-0.344\vec{i} + 6.23\vec{j} + 2.511\vec{k})$.

(c) As the next journey starts 1 hour (3600 seconds) after the first journey
$$\vec{r} = (310\vec{i} + 1630\vec{j}) + (t - 3600)(-0.344\vec{i} + 6.23\vec{j} + 2.511\vec{k})$$
$$= (310\vec{i} + 1630\vec{j}) + 1240\vec{i} - 22440\vec{j} - 9040\vec{k} + t(-0.344\vec{i} + 6.233\vec{j} + 2.511\vec{k})$$
$$= (1550\vec{i} - 20810\vec{j} - 9040\vec{k}) + t(-0.344\vec{i} + 6.233\vec{j} + 2.511\vec{k})$$

(d) At 9.15 am, $t = 4500 \Rightarrow \vec{r} = (1550\vec{i} - 20810\vec{j} - 9040\vec{k}) + 4500(-0.344\vec{i} + 6.233\vec{j} + 2.511\vec{k})$
$$= (1550\vec{i} - 20810\vec{j} - 9040\vec{k}) + (-1550\vec{i} + 28050\vec{j} + 11300\vec{k})$$
$$= 0\vec{i} + 7240\vec{j} + 2260\vec{k},$$
which is the position vector of the upper terminus.

Solutions to Unit 6 Exercises

Exercise 6.1

For questions 1 – 4 the frequencies will depend on the random numbers which are generated. The column for tally marks has been omitted.

1.

Score	1	2	3	4	5	6
Frequency	16	11	6	6	6	15

2.

Score	2	3	4	5	6	7	8	9	10	11	12
Frequency	0	1	5	8	12	13	7	6	4	4	0

3.

Number of Tosses	1	2	3	4	5	6	7	8
Frequency	11	10	7	0	1	0	0	1

4.

Number	0-9	10-19	20-29	30-39	40-49
Frequency	7	10	7	8	9
Number	50-59	60-69	70-79	80-89	90-99
Frequency	6	11	5	9	8

5.

Class Interval	Tally Marks	Frequency
300 – 349	111	3
350 – 399	11	2
400 – 449	11111	5
450 – 499	11111 111	8
500 – 549	11111 111	8
550 – 599	11111 1111	9
600 – 649	11111 11	7
650 – 699	11	2
700 – 749	11111	5
750 – 799		0
800 – 850	1	1

6.

Time	0.5-1	1-1.5	1.5-2	2-2.5	2.5-3	3-3.5
Frequency	3	10	19	5	6	3
Time	3.5-4	4-4.5	4.5-5	5-5.5	5.5-6	
Frequency	0	1	2	0	1	

Exercise 6.2

1. (a)

Number	170-174	175-179	180-184	185-189	190-194
Frequency	1	6	11	10	5

(b)

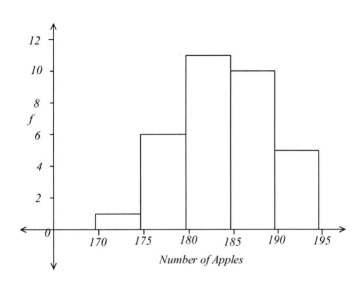

2. The frequencies and hence the shape of the frequency histogram will depend on the random numbers generated.

(a)

Class Interval	0-9	10-19	20-29	30-39	40-49
Frequency	8	5	11	19	9
Class Interval	50-59	60-69	70-79	80-89	90-99
Frequency	9	16	6	14	3

(b)

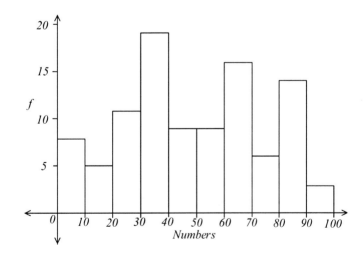

(c) As the random numbers have an equal probability of being any integer between 0 and 99, one might expect the height of the rectangles of the frequency histogram to be roughly equal. This frequency histogram does not conform to this shape. If many more random numbers were generated, then the heights of the rectangles might be more equal.

3. (a)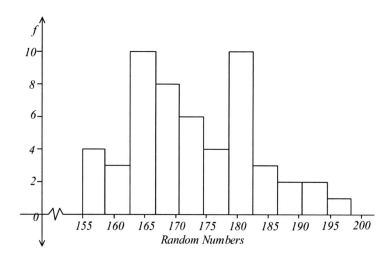

(b)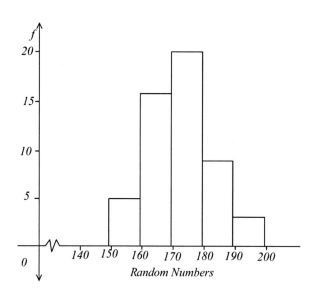

Exercise 6.3

1.

Grade	Cumulative Frequency
1	2
2	7
3	14
4	25
5	39
6	45
7	47

2. (a)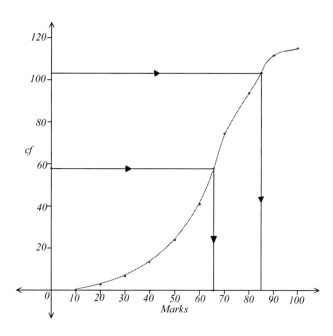

(b) The construction lines for part (b) are shown on the cumulative frequency curve for part (a).

   (i) $0.90 \times 115 = 103.5$. Therefore the 90th percentile is 85.

   (ii) $0.5 \times 115 = 57.5$. Therefore the 50th percentile is 66.

3. (a)   (b)

Upper Class Boundary	Cum Freq
0	0
9.95	2
19.95	13
29.95	30
39.95	37
49.95	48
59.95	55
69.95	61
79.95	70
89.95	74
99.95	82
109.95	87

4. The grouping of the data will depend on the class intervals chosen. This will also affect the cumulative frequency table and curve.

(a)

Class Interval	Frequency
37.95 – 38.25	2
38.25 – 38.55	1
38.55 – 38.85	6
38.85 – 39.15	16
39.15 – 39.45	15
39.45 – 39.75	6

(b)

Upper Class Boundary	Cumulative Frequency
≤ 37.95	0
≤ 38.25	2
≤ 38.55	3
≤ 38.85	9
≤ 39.15	25
≤ 39.45	40
≤ 39.75	46

(c)

(d) The construction lines are shown on the cumulative frequency curve drawn for part (c).

(i) $0.9 \times 46 = 41.4$. Therefore, 90th percentile is $39.52°C$.

(ii) $1 - 0.6 = 0.4$, $0.4 \times 46 = 18.4$. Therefore, the temperature is $39.04°C$.

These values will vary slightly depending on the way the curve is drawn.

5. (a)

Upper Class Limit	Cumulative Frequency
0	0
≤ 3.0	74
≤ 3.5	129
≤ 4.0	168
≤ 4.5	179
≤ 5.0	186
≤ 6.0	189
≤ 7.0	190

(b)

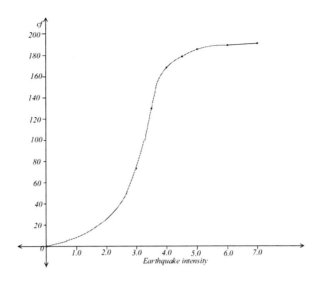

(c) (i) $0.95 \times 190 = 180.5$. Therefore, the 95th percentile is 4.6.

(ii) $0.6 \times 190 = 114$. Therefore, the 60th percentile is 3.3.

(iii) $0.9 \times 190 = 171$. Therefore, the 90th percentile is 4.1.

(d) $0.25 \times 190 = 47.5$. Therefore, the lower quartile is 2.7.

$0.75 \times 190 = 142.5$. Therefore, the upper quartile is 3.7.

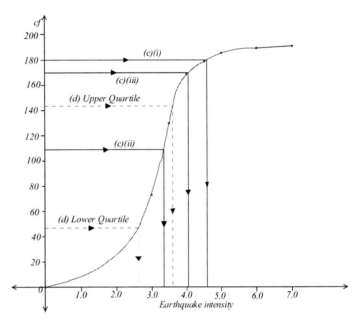

Exercise 6.4

1. (a) 17.0   (b) 0.509   (c) 1.15×10^{-8}

2.

x_i	f_i	$f_i x_i$
0	8	0
1	5	5
2	4	8
3	4	12
4	0	0
5	1	5
6	0	0
7	1	7

$$\bar{x} = \frac{\sum f_i x_i}{\sum f_i} = \frac{37}{23} = 1.61$$

3. $\bar{x} = \dfrac{\sum f_i x_i}{\sum f_i} = \dfrac{362}{45} = 8.04$

Mid-Interval Value, x_i	Frequency, f_i	$f_i x_i$
4	3	12
6	12	72
8	15	120
10	11	110
12	4	48

4. (a) $\bar{x} = 3.63$ (b) $\bar{x} = 46.1$ (c) $\bar{x} = 0.397$

5. 1.97mg

6. (a) $\bar{x} = 4.7608$

   (b) This is most easily done by sorting the data in ascending order in your graphing calculator.

   (c) $\bar{x} = 4.76$

   (d) The mean calculated in part (c) assumes that the data in each class interval are evenly distributed across the interval. In practice, this is rarely true. For example, the three data points in the class interval 0 – 1 are 0.036, 0.298 and 0.902, which are spread quite well across the interval but not very evenly. However, it can be seen from the small difference between the two means that the assumption is generally valid.

Class Interval	Mid-Interval Value	Freq
0 – 1	0.5	3
1 – 2	1.5	5
2 – 3	2.5	6
3 – 4	3.5	10
4 – 5	4.5	4
5 – 6	5.5	5
6 – 7	6.5	6
7 – 8	7.5	3
8 – 9	8.5	4
9 – 10	9.5	4

Exercise 6.5

1. (a) 42   (b) 1.935   (c) 589

2. (a) median is 6, mode is 5   (b) median is 3, mode is 2

3. (a)

Upper Boundary Limit	Cumulative Frequency
0	0
20	28
30	87
40	124
50	148
60	159
70	172

(b)

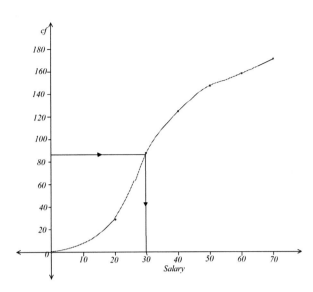

(c) Median salary is $30000

(d) Modal class of salaries is from $20000 to $30000

4.   Median mark is 66

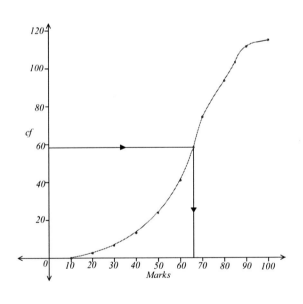

5. 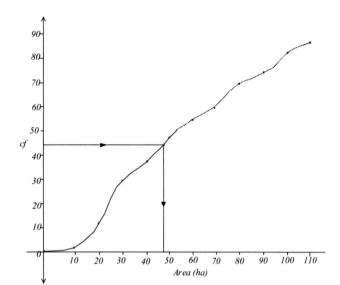   Median is 47ha.

Exercise 6.6

1. (a) The median is 385, the lower quartile is 380 and the upper quartile is 396.5. Therefore, the interquartile range is 16.5.

   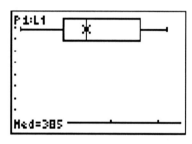

   (b) 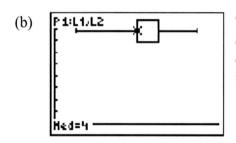 The median is 4, the lower quartile is 4 and the upper quartile is 5. Therefore, the interquartile range is 1.

2. (a)

Upper Boundary Limit	Cumulative Frequency
−40°	0
−35°	5
−30°	16
−25°	42
−20°	79
−15°	97
−10°	100

106

(b) The lower quartile is $-27.5°C$ and the upper quartile is $-20.5°C$. Therefore, the interquartile range is $7°C$.

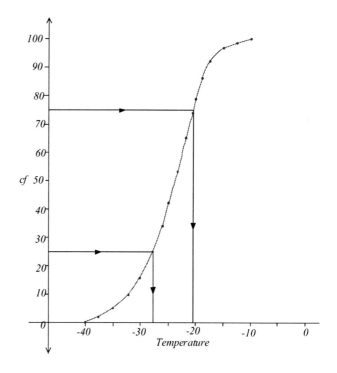

3. (a)

x_i	$x_i - \bar{x}$	$(x_i - \bar{x})^2$
8.5	0.5	0.25
7.8	-0.2	0.04
8.1	0.1	0.01
7.9	-0.1	0.01
7.7	-0.3	0.09
8.0	0	0

$$\bar{x} = \frac{8.5+7.8+8.1+7.9+7.7+8.0}{6} = 8,$$

$$\sum(x_i - \bar{x})^2 = 0.4$$

Therefore, $s = \sqrt{\frac{1}{6} \times 0.4} = 0.258$.

(b) $\bar{x} = \dfrac{123+134+109+115+117+124+130+122+125+111}{10}$

$= 121$

$\sum(x_i - \bar{x})^2 = 576$. Therefore $s = \sqrt{\dfrac{1}{10} \times 576} = 7.59$

x_i	$x_i - \bar{x}$	$(x_i - \bar{x})^2$
123	2	4
134	13	169
109	-12	144
115	-6	36
117	-4	16
124	3	9
130	9	81
122	1	1
125	4	16
111	10	100

4.

x_i	f_i	f_ix_i	$f_i(x_i-\bar{x})^2$
10	2	20	800
20	11	220	1100
30	15	450	0
40	9	360	900
50	3	150	1200

$\bar{x} = \dfrac{1200}{40} = 30$, $\sum f_i(x_i - \bar{x})^2 = 4000$. Therefore, $s = \sqrt{\dfrac{1}{40} \times 4000} = 10$.

5.   (a) $s = 0.0762$

       (b)

Class Interval	Frequency
0.8 – 0.9	3
0.9 – 1.0	6
1.0 – 1.1	2
1.1 – 1.2	1

          Standard deviation is 0.0862

       (c) In grouping the data, it is assumed that all data points in a class interval are located at the center of that interval. For example, all three data points in the interval 0.8 – 0.9 are assumed to be at 0.85, whereas, in fact, they are 0.869, 0.876, 0.887. Therefore, the answer to part (b) differs slightly from the answer of part (a).

6.   (a) (i) $s = 2.29$ (ii) $s = 0.0597$ (b) (i) $s = 2.29$ (ii) $s = \dfrac{1}{10}(0.0597) = 0.00597$

7.   (a) The standard deviation of the English exam marks is 16.5.
          The standard deviation of the Mathematics exam marks is 10.4.

       (b) By looking at the frequency table, you can see that there are more low marks and more high marks in the English exam than in the Math exam. Therefore, the mark distribution of the English exam is spread more widely than the Math exam, and so you would expect the standard deviation of the English exam marks to be higher than the standard deviation of the Math exam marks.

8. $s_A = 9.50$, $s_B = 4.72$. By looking at the frequency table, you can see that, in Pluvaria, there are more years of low rainfall and more years of high rainfall than in Torrentia. Therefore, you would expect that the standard deviation of rainfall in Pluvaria to be higher than in Torrentia, where the rainfall is less widely distributed.

Exercise 6.7

1. (a) $P(A) = \frac{1}{6}$, $P(B) = \frac{1}{6}$ (i) not exhaustive   (ii) mutually exclusive

   (b) $P(A) = \frac{1}{6}$, $P(B) = \frac{1}{2}$ (i) not exhaustive   (ii) not mutually exclusive

   (c) $P(A) = \frac{1}{2}$, $P(B) = \frac{1}{2}$ (i) exhaustive   (ii) mutually exclusive

   (d) $P(A) = \frac{2}{3}$, $P(B) = \frac{1}{6}$ (i) not exhaustive   (ii) mutually exclusive

   (e) $P(A) = \frac{1}{2}$, $P(B) = \frac{2}{3}$ (i) exhaustive   (ii) not mutually exclusive

2.

	1	2	3	4	5	6
1	2	3	4	5	6	7
2	3	4	5	6	7	8
3	4	5	6	7	8	9
4	5	6	7	8	9	10
5	6	7	8	9	10	11
6	7	8	9	10	11	12

   (a) $P(A) = \frac{5}{18}$, $P(B) = \frac{1}{9}$ (i) not exhaustive   (ii) not mutually exclusive

   (b) $P(A) = \frac{11}{36}$, $P(B) = \frac{1}{36}$ (i) not exhaustive   (ii) mutually exclusive

   (c) $P(A) = \frac{1}{6}$, $P(B) = \frac{5}{6}$ (i) exhaustive   (ii) mutually exclusive

   (d) $P(A) = \frac{1}{36}$, $P(B) = \frac{1}{36}$ (i) not exhaustive   (ii) not mutually exclusive

   (e) $P(A) = 1$, $P(B) = \frac{1}{6}$ (i) exhaustive   (ii) not mutually exclusive

3. (a) $P(W) = 1 - (0.2 + 0.5) = 0.3$

   (b) $P(A \cup W) = P(A) + P(W)$ because A and W are mutually exclusive. Therefore, the probability is $0.2 + 0.3 = 0.5$.

4. (a) $\frac{2}{5}$ (b) $\frac{3}{10}$ (c) $\frac{2}{5}$ (d) $\frac{3}{5}$

5. (a) 0.19 (b) 0.26

6. (a) $\frac{4}{25}$ (b) $\frac{9}{25}$

	R	R	G	G	B
R	(RR)	(RR)	(RG)	(RG)	(RB)
R	(RR)	(RR)	(RG)	(RG)	(RB)
G	(GR)	(GR)	(GG)	(GG)	(GB)
G	(GR)	(GR)	(GG)	(GG)	(GB)
B	(BR)	(BR)	(BG)	(BG)	(BB)

7. (a) $\frac{3}{10}$ (b) $\frac{1}{10}$ (c) $\frac{2}{5}$

8. (a) $0.92^2 = 0.8464$ (b) $0.92 \times 0.08 + 0.08 \times 0.92 = 0.1472$

	B	S
B	(BB)	(BS)
S	(SB)	(SS)

9. (a) $\frac{1}{10}$ (b) $\frac{3}{10}$

Exercise 6.8

1. (a) $\frac{5}{8}$ (b) $\frac{3}{8}$ (c) $\frac{3}{4}$ (d) $\frac{1}{4}$

2. (a) $\frac{1}{2}$ (b) $\frac{1}{4}$ (c) $\frac{5}{8}$ (d) $\frac{5}{8}$

3. (a)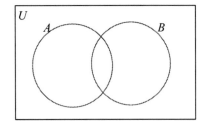

(b) $n(A \cap B) = 50 - 18 = 32$
$n(A \cup B) = n(A) + n(B) - n(A \cap B)$.
$\Rightarrow 32 = 20 + 20 - n(A \cap B) \Rightarrow n(A \cap B) = 8$

(c) $P(A \cap B) = \frac{4}{25}$

4. (a)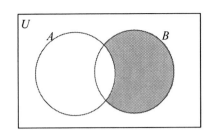

   (b) $n(A' \cap B) = n(B) - n(A \cap B)$

   (c) (i) 10 (ii) $\frac{1}{8}$

   (d) A and B are not mutually exclusive because $n(A \cap B) \neq 0$. (Alternatively, because $P(A \cap B) \neq 0$.)

5. (a) $P(A \cup B) = P(A) + P(B) - P(A \cap B) \Rightarrow 0.95 = 0.62 + 0.51 - P(A \cap B)$
   $\Rightarrow P(A \cap B) = 0.18$

   (b) $P(A \cup B') = 1 - P(B) + P(A \cap B) = 1 - 0.51 + 0.18 = 0.67$

6. (a) $P(A \cup B) = P(A) + P(B) - P(A \cap B) = \frac{1}{3} + \frac{3}{4} - \frac{1}{5} = \frac{53}{60}$

   (b) $P(A' \cup B') = 1 - P(A \cap B) = 1 - \frac{1}{5} = \frac{4}{5}$

7. (a) $\frac{3}{4} = \frac{1}{2} + P(B) - \frac{1}{6} \Rightarrow P(B) = \frac{3}{4} + \frac{1}{6} - \frac{1}{2} = \frac{5}{12}$

   (b) $P(A' \cap B) = P(B) - P(A \cap B) = \frac{5}{12} - \frac{1}{6} = \frac{1}{4}$

8. $P(A' \cup B') = 1 - P(A \cap B) = 0.8 \Rightarrow P(A \cap B) = 0.2 \Rightarrow P(A \cup B) = 0.5 + 0.4 - 0.2 = 0.7$

9. $P(A \cap B') + P(A' \cap B) + P(A \cap B) = P(A \cup B) \Rightarrow 0.23 + 0.42 + P(A \cap B) = 0.93$
   $\Rightarrow p(A \cap B) = 0.28$

10. The probability that a randomly chosen student studies math not physics is $\frac{80}{200} = \frac{2}{5}$

    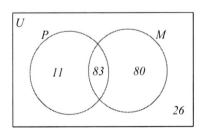

11. $P(L) = 0.16$, $P(B) = 0.28$, $P(L \cap B) = 0.08$,
    $P(L \cup B) = 0.16 + 0.28 - 0.08 = 0.36$.

Exercise 6.9

1. (a) $\dfrac{39}{125}$ (b) $\dfrac{31}{57}$ (c) $\dfrac{1}{3}$

2. (a) 0.52   (b) 0.33   (c) 0.68

3. (a) $\dfrac{11}{72}$ (b) $\dfrac{7}{18}$

4. (a) $P(A|B) = \dfrac{P(A \cap B)}{P(B)} = \dfrac{1}{4}$

    (b) $P(A \cap B') = P(A) - P(A \cap B) \Rightarrow P(A|B') = \dfrac{P(A \cap B')}{P(B')} = \dfrac{P(A) - P(A \cap B)}{P(B')} = \dfrac{5}{8}$

5. (a) $P(A|B) = \dfrac{P(A \cap B)}{P(B)} = \dfrac{3}{4}$ (b) $P(B|A) = \dfrac{P(B \cap A)}{P(A)} = \dfrac{P(A \cap B)}{P(A)} = \dfrac{5}{8}$

6. (a) $P(P|Q) = \dfrac{P(P \cap Q)}{P(Q)} \Rightarrow \dfrac{3}{7} = \dfrac{P(P \cap Q)}{\frac{2}{3}} \Rightarrow P(P \cap Q) = \dfrac{2}{7}$

    (b) $P(Q|P) = \dfrac{P(P \cap Q)}{P(P)} \Rightarrow P(P) = \dfrac{P(P \cap Q)}{P(Q|P)} = \dfrac{5}{14}$

7.

    (a) $P(A \cap B) = \dfrac{3}{8} \times \dfrac{2}{3} = \dfrac{1}{4}$ (b) $P(A' \cap B') = \dfrac{5}{8} \times \dfrac{3}{5} = \dfrac{3}{8}$

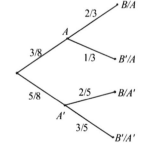

8. (a) $P(B|A) = \dfrac{P(A \cap B)}{P(A)} = \dfrac{0.36}{0.48} = 0.75$

    (b) $P(B'|A') = \dfrac{P(A' \cap B')}{P(A')} = \dfrac{0.26}{0.52} = 0.5$

    (c) $P(B) = P(A) \times P(B|A) + P(A') \times P(B|A') = 0.62$

    (d) $P(A|B) = \dfrac{P(A \cap B)}{P(B)} = \dfrac{0.36}{0.62} = 0.581$

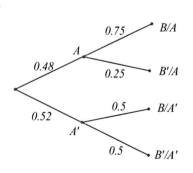

9. $P(B|A) = \dfrac{P(A \cap B)}{P(A)} = \dfrac{2}{5}$, $P(A|B) = \dfrac{P(A \cap B)}{P(B)} \Rightarrow P(A \cap B) = \dfrac{1}{3} \times \dfrac{4}{5} = \dfrac{4}{15} \Rightarrow P(A) = \dfrac{4/15}{2/5} = \dfrac{2}{3}$

10. (a)

(b) $P(D) = P(D/V) \times P(V) + P(D/V') \times P(V')$
$= 0.155$

(c) $P(V \cap D) = P(V) \times P(D|V) = 0.015$

Therefore, $P(V|D) = \dfrac{P(V \cap D)}{P(D)} = 0.0968$

11. (a) $P(A) = 0.7$, $P(D|A') = 0.05$, $P(D|A) = 0.01$

(b) $P(D) = P(D|A) \times P(A) + P(D|A') \times P(A') = 0.022$ and $P(A' \cap D) = \dfrac{0.015}{0.022} = 0.682$

12. (a) $P(A) = \dfrac{1}{100}$, $P(B|A) = 1$, $P(B|A') = \dfrac{1}{1024}$

(b) $P(B) = P(B|A) \times P(A) + P(B|A') \times P(A') = 0.0109668$

$\Rightarrow P(A|B) = \dfrac{P(A \cap B)}{P(B)} = \dfrac{0.01 \times 1}{0.0109668} = 0.912$

Exercise 6.10

1. (a) $P(A \cap B) = \dfrac{1}{36}$ (b) $P(A \cup B) = \dfrac{23}{72}$

2. (a) $P(B) = \dfrac{11}{16}$ (b) $P(A \cap B) = \dfrac{33}{80}$

3. $P(A \cap B) = 0.3$, $P(A) p(B) = 0.3$. Therefore A and B are independent events.

4. $P(B) = 0.6$, $P(A) P(B) = 0.45 \times 0.6 = 0.27 = P(A \cap B)$. So A and B are independent events.

5. (a) Let $P(Q) = x$, (a) $0.8 = x + 3x - 3x^2 \Rightarrow 3x^2 - 4x + 0.8 = 0 \Rightarrow x = 0.2450, 1.0883$ but $|x| \leq 1$ so that $P(Q) = 0.245$.

(b) $P(P \cup Q) = 1 \Rightarrow 1 = 4x - 3x^2 \Rightarrow (3x-1)(x-1) = 0 \Rightarrow P(Q) = \dfrac{1}{3}$. (If $x = 1$, then $P(P) = 3$.)

6. (a) $\frac{1}{8}$ (b) $\frac{3}{8}$ (c) $\frac{5}{8}$ (d) $\frac{3}{8}$

7. (a) $\frac{8}{81}$ (b) $\frac{16}{81}$ (c) $\frac{1}{9}$ (d) $\frac{29}{81}$

8. Letters in common are M, I and S. Then the probability that Ms are chosen is $\frac{1}{11} \times \frac{1}{8} = \frac{1}{88}$. The probability that Is are chosen is $\frac{4}{11} \times \frac{2}{8} = \frac{1}{11}$. The probability that Ss are chosen is $\frac{4}{11} \times \frac{2}{8} = \frac{1}{11}$. Therefore, the probability that the letters chosen are the same is $\frac{1}{88} + \frac{1}{11} + \frac{1}{11} = \frac{17}{88} \approx 0.193$, and so the probability that the same letters are chosen is 19.3%.

9. (a) $P(TTTH) = \frac{1}{16}$

    (b) $P(n \text{ tails}) = \left(\frac{1}{2}\right)^n \leq 0.1 \Rightarrow \log\left(\frac{1}{2}\right)^n \leq -1 \Rightarrow n \geq 3.32$. Therefore, $n = 4$.

10. The probability of not scoring on n throws is $\left(\frac{1}{2}\right)^n$. This probability must be less than 0.01. Therefore, $\left(\frac{1}{2}\right)^n \leq 0.01 \Rightarrow \log\left(\frac{1}{2}\right)^n \leq -2 \Rightarrow n \geq \frac{-2}{\log 0.5} = 6.64$ and the player would need to take 7 free throws.

11. (a) $\left(1 - \frac{1}{6}\right)^n \leq 0.05 \Rightarrow n \log\left(\frac{5}{6}\right) \leq \log 0.05 \Rightarrow n \geq 16.43$. Therefore, the die should be thrown 17 times.

    (b) If die is to be thrown at least 10 times, we require 9 non-sixes of which the probability is $\left(\frac{5}{6}\right)^9 = 0.1938$. Therefore, the probability is 0.194.

12. Let C be the event: agreement on chemical weapons. Let N be the event: agreement on nuclear weapons. Then $P(C) = 0.5$, $P(N) = 0.7$ and $P(C' \cap N') = 0.2\sqrt{2}$. Therefore, as $P(C' \cap N') = 1 - P(C \cup N)$, $P(C \cup N) = 0.8 \Rightarrow P(C \cap N) = 0.5 + 0.7 - 0.8 = 0.4$. Therefore, the probability of agreement on both chemical and nuclear weapons is 0.4.
As $P(C) \times P(N) = 0.35 \neq 0.4$, the two issues are not treated independently.

Exercise 6.11

1. (a) $c = 0.1$ (b) $c = 0.05$ (c) $c = 0.5$

2. (a)

	1	2	3	4	5	6
1	0	1	2	3	4	5
2	1	0	1	2	3	4
3	2	1	0	1	2	3
4	3	2	1	0	1	2
5	4	3	2	1	0	1
6	5	4	3	2	1	0

(b)

X	0	1	2	3	4	5
$P(X = x)$	1/6	5/18	2/9	1/6	1/9	1/18

3. (a)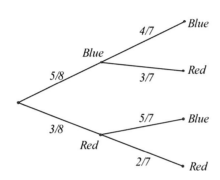

(b)

X	0	1	2
$P(X = x)$	$\frac{3}{28}$	$\frac{15}{28}$	$\frac{5}{14}$

4. (a)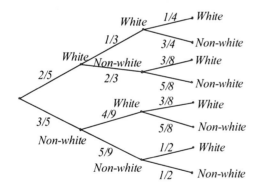

(b)

X	0	1	2	3
$P(X = x)$	$\frac{1}{6}$	$\frac{1}{2}$	$\frac{3}{10}$	$\frac{1}{30}$

(c) $E(X) = 1.2$

5. (a) The probability of 'TTH' is $\left(\dfrac{1}{2}\right)^3 = \dfrac{1}{8}$.

(b) 3 points can be scored either by 'HTT' or by 'THH'.

Therefore, $P(X = 3) = \left(\dfrac{1}{2}\right)^3 + \left(\dfrac{1}{2}\right)^3 = \dfrac{1}{8} + \dfrac{1}{8} = \dfrac{1}{4}$.

(c)

X	0	1	2	3	4	5	6
$P(X = x)$	$\frac{1}{8}$	$\frac{1}{8}$	$\frac{1}{8}$	$\frac{1}{4}$	$\frac{1}{8}$	$\frac{1}{8}$	$\frac{1}{8}$

(d) $E(X) = 3$

6. (a) $2, $10, $50 and $1000 are prizes but it costs $1 to play. Therefore, if the player loses $X = -1$, if the player wins $X = 1, 9, 49, 999$, depending on the extent of his/her winning.

   (b)

x	-1	1	9	49	999
$P(X = x)$	0.8445	0.125	0.027	0.003375	0.000125

   (c) $E(X) = -0.18625$ so that the owner makes $186.25.

7. (a) 4.29   (b) $P(X = x) = \frac{1}{10}$ for all $x \in \{0,1,2, \ldots, 9\} \Rightarrow E(X) = 4.5$

8. (a) Let X be the event "the lavatory is occupied"

X	0	1	2	3
$P(X = x)$	0.216	0.432	0.288	0.064

   (b) $E(X) = 1.2$

9. (a) The 13 possible PIN numbers are: 106, 610, 160, 601, 133, 313, 331, 124, 142, 214, 241, 412, 421. Let X be the event "the correct PIN is guessed", then $P(X = 1) = \frac{1}{13}$,

   $P(X = 2) = \frac{12}{13} \times \frac{1}{12} = \frac{1}{13}$, $P(X = 3) = \frac{12}{13} \times \frac{11}{12} \times \frac{1}{11} = \frac{1}{13}$ and so on. Therefore,

   $P(X = x) = \frac{1}{13}$ for all values of x.

   (b) $E(X) = \frac{1}{13}(1 + 2 + 3 + \ldots + 13) = 7$

Exercise 6.12

1. (a) $\frac{1}{4}$ (b) $\frac{5}{16}$

2. (a) $\frac{1}{64}$ (b) $\frac{5}{16}$ (c) $\frac{3}{32}$

3. (a) 0.372   (b) 0.260   (c) 0.135

4. (a) 0.478   (b) 0.124   (c) 0.0230

Exercise 6.13

1. (a) 0.00412   (b) 0.165

2. 0.313

3. 0.121

4. (a) 0.277   (b) 0.231   (c) 0.127

5. (a) 0.665   (b) 0.619   (c) 0.597   (d) 0.584

6. Let X be the event 'a person who has booked a ticket fails to take up the booking'.
$n = 50$, $p = \frac{1}{16}$, $P(X \leq 1) = 0.1719$. Therefore, the probability that there will not be enough seats is 0.172.

7. (a) 0.133   (b) 0.311

8. (a) 0.123   (b) 0.00856   (c) 0.991   (d) 0.945   (e) 0.900

9. (a) $P(X \geq 1) = 1 - P(X = 0) = 1 - (0.992)^7 = 0.05467 \approx 5\%$

(b) (i) 0.799   (ii) 0.185   (iii) 0.0160

Exercise 6.14

1. (a) $E(X) = 25$ (b) $E(X) = 10$

2. (i) (a) 1   (b) 0.402   (ii) (a) 10   (b) 0.137

3. (a) Let X be the number of defective components. Then $X \sim B(120, 0.06)$
$E(X) = np = 120 \times 0.06 = 7.2$
(b) Let Y be the number of families that are willing to host a refugee.
Then $Y \sim B(25\,000, 0.35)$ and $E(Y) = 25000 \times 0.35 = 8750$

4. (a) $np = 15$, $p = 0.3 \Rightarrow n = \frac{15}{0.3} = 50$ (b) $8p = 5 \Rightarrow p = \frac{5}{8}$

5. $E(X) = E(Y) \Rightarrow 10p = 30p^2 \Rightarrow 3p^2 - p = 0 \Rightarrow p(3p - 1) = 0 \Rightarrow p = 0,\ 3p - 1 = 0$
so the non-zero value of p is $\frac{1}{3}$.

6. (a) $\binom{3}{2}(1-p)p^2 = 0.2 \Rightarrow 3p^2 - 3p^3 = 0.2 \Rightarrow 15p^3 - 15p^2 + 1 = 0$

(b) The solutions are -0.233, 0.311, 0.921.

(c) As $0 \leq p \leq 1$, the only valid solutions are 0.311 and 0.921

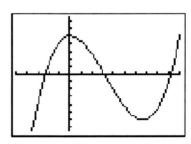

7. (a) $n = 3$. This should be evident from the observation that $X \in \{0, 1, 2, 3\}$.

(b) $E(X) = 1 \times \dfrac{48}{125} + 2 \times \dfrac{12}{125} + 3 \times \dfrac{1}{125} = \dfrac{48 + 24 + 3}{125} = \dfrac{3}{5} = 0.6$

(c) $E(X) = np \Rightarrow 0.6 = 3p \Rightarrow p = \dfrac{0.6}{3} = 0.2 = \dfrac{1}{5}$.

8. (a) $\bar{x} = \dfrac{1}{n} \sum_{i=1}^{N} f_i x_i = \dfrac{0 \times 16 + 1 \times 33 + 2 \times 31 + 3 \times 18 + 4 \times 14 + 5 \times 7 + 6 \times 1}{120} = 2.05$

(b) Let X be a random variable representing the number of red flowers in a packet of six.
$E(X) = np \Rightarrow 2.05 = 6p \Rightarrow p = 0.342$

(c) $P(X = 0) = (1 - 0.342)^6 = 0.0812$ so number of samples with 0 red flowers is $0.0812 \times 120 \approx 10$

$P(X = 1) = \binom{6}{1} \times 0.658^5 \times 0.342 = 0.253$ and number of samples with 1 red flower is $0.253 \times 120 \approx 30$.

The other frequencies are calculated in a similar way and the frequency distribution is

Number of red flowers, x_i	0	1	2	3	4	5	6
Number of samples with x_i red flowers	10	30	39	27	11	2	0

Exercise 6.15

1. (a) 0.9893 (b) 0.6368 (c) 0.9228 (d) 0.1894 (e) 0.0034 (f) 0.3181

   (g) 0.2981 (h) 0.9649 (i) 0.0195 (j) 0.8186 (k) 0.5762 (l) 0.0789

2. (a) $P(|Z| < 1) = P(-1 < Z < 1) = 0.6827$

   (b) $P(|Z| < 2.5) = P(-2.5 < Z < 2.5) = 0.9876$

   (c) $P(|Z| > 3) = 1 - P(|Z| < 3) = 1 - P(-3 < Z < 3) = 1 - 0.997300 = 0.0027$

   (d) $P(|Z - 1| < 0.1) = P(-0.1 < Z - 1 < 0.1) = P(0.9 < Z < 1.1) = 0.0484$

   (e) $P(|Z - 0.5| > 1.5) = 1 - P(|Z - 0.5| < 1.5) = 1 - P(-1.5 < Z - 0.5 < 1.5) = 1 - P(-1 < Z < 2)$
   $= 1 - 0.8186 = 0.1814$

Exercise 6.16

1. (i) (a) 0.9554 (b) 0.2119 (c) 0.6612

8. Let X be the diameter of a randomly chosen washer.

   For the original machine $P(X > 3.758) = 0.0098153$ so for the upgraded machine we also require that $P(X > 3.758) = 0.0098153$, therefore $P(X \leq 3.758) = 0.990185$ but $P(Z < z_1) = 0.990185 \Rightarrow z_1 = 2.33335$ and as $Z = \dfrac{X - \mu}{\sigma}$, we have $2.33335 = \dfrac{3.758 - \mu}{0.002}$.

   Therefore $\mu = 3.758 - 2.33335 \times 0.002 = 3.75333$.

   So, on the upgraded machine, the mean should be set at 3.753cm. Although the question does not request a solution to three decimal places the mean diameter of the washers under the original setting was given to three decimal places so it makes sense to give the new mean diameter of the washers to the same accuracy.

Solutions to Unit 7 Exercises

Exercise 7.1

In each case, the graph of the function is drawn and the tangents found at regular intervals. The gradients of these tangents are then entered into lists and the suggested regression carried out.

1.

x	0	1	2	3	4	5	6
m	0	3	12	27	48	75	108

$f'(x) = 3x^2$

2.

x	-3	-2	-1	0	1	2
m	-13	-7	-1	5	11	17

$f'(x) = 6x + 5$

3.

x	-2	-1	0	1	2
m	-32	-4	0	4	32

$f'(x) = 4x^3$

4.

x	1	2	3	4	5
m	0.5	0.3535	0.2887	0.25	0.2236

$f'(x) = 0.5x^{-0.5}$

5. $f'(x) = 3ax^2 + 2bx + c$

Exercise 7.2

1. $f'(x) = \lim_{h \to 0} \left(\dfrac{f(x+h) - f(x)}{h} \right) \Rightarrow f'(x) = \lim_{h \to 0} \left(\dfrac{(x+h)^2 + 3(x+h) - (x^2 + 3x)}{h} \right)$

$= \lim_{h \to 0} \left(\dfrac{x^2 + 2xh + h^2 + 3x + 3h - x^2 - 3x}{h} \right) = \lim_{h \to 0} \left(\dfrac{2xh + h^2 + 3h}{h} \right)$

$\Rightarrow f'(x) = \lim_{h \to 0} \left(\dfrac{h(2x + h + 3)}{h} \right) = \lim_{h \to 0} (2x + 3 + h) \Rightarrow f'(x) = 2x + 3$

2. $f'(x) = \lim\limits_{h \to 0}\left(\dfrac{f(x+h)-f(x)}{h}\right) = \lim\limits_{h \to 0}\left(\dfrac{(3-2(x+h))-(3-2x)}{h}\right)$

$\Rightarrow f'(x) = \lim\limits_{h \to 0}\left(\dfrac{-2h}{h}\right) = -2$

3. $f'(x) = \lim\limits_{h \to 0}\left(\dfrac{((x+h)^3 + 4(x+h)^2) - (x^3 + 4x^2)}{h}\right)$

$\Rightarrow f'(x) = \lim\limits_{h \to 0}\left(\dfrac{(x^3 + 3x^2h + 3xh^2 + h^3 + 4(x^2 + 2xh + h^2)) - (x^3 + 4x^2)}{h}\right)$

$\Rightarrow f'(x) = \lim\limits_{h \to 0}\left(\dfrac{3x^2h + 3xh^2 + h^3 + 8xh + 4h^2}{h}\right) = \lim\limits_{h \to 0}\left(\dfrac{h(3x^2 + 3xh + h^2 + 8x + 4h)}{h}\right)$

$\Rightarrow f'(x) = \lim\limits_{h \to 0}(3x^2 + 3xh + h^2 + 8x + 4h) = 3x^2 + 8x$

Exercise 7.3

1. (a) $f'(x) = 7x^6$ (b) $f'(x) = 45x^4$ (c) $f'(x) = 1 - 9x^2$ (d) $f'(x) = 6x$

   (e) $f'(x) = 8x^3 + 6x - 6$ (f) $f'(x) = \dfrac{5}{3}x^4 - \dfrac{6}{5}x^2 - \dfrac{1}{2}$ (g) $f'(x) = x^2$

   (h) $f'(x) = 1 + \dfrac{3}{x^2}$ (i) $f'(x) = 5x^4 + \dfrac{5}{x^6}$ (j) $f'(x) = (2k+1)x^{2k}$

2. (a) $g'(x) = \dfrac{3}{2\sqrt{x}}$ (b) $g'(x) = 4x^3 - \dfrac{1}{x^2}$

   (c) $g'(x) = -\dfrac{3}{x^4} + \dfrac{2}{x^2}$ (d) $g'(x) = \dfrac{1}{2\sqrt{x}} - \dfrac{1}{2\sqrt{x^3}}$

3. (a) $f'(x) = 4x^3 \Rightarrow f'(2) = 32$ (b) $f'(x) = 4x^3 \Rightarrow f'(2) = 32$

   (c) $f'(x) = 2x - 9 \Rightarrow f'(2) = -5$ (d) $f'(x) = -\dfrac{2}{3}x^{-3} \Rightarrow f'(2) = -\dfrac{1}{12}$

   (e) $f'(x) = 2x^{-\frac{1}{2}} \Rightarrow f'(2) = \sqrt{2}$

Exercise 7.4

1. (a) $\dfrac{dy}{dx} = 7x^6$ (b) $\dfrac{dy}{dx} = 24x^5$ (c) $\dfrac{dy}{dx} = -21x^2$ (d) $\dfrac{dy}{dx} = 48x^3 - 21x^2 - 6x + 5$

   (e) $y = 4x^{-3} + 3x^{-2} - \dfrac{2}{3}x^{-1} \Rightarrow \dfrac{dy}{dx} = -12x^{-4} - 6x^{-3} + \dfrac{2}{3}x^{-2} = -\dfrac{12}{x^4} - \dfrac{6}{x^3} + \dfrac{2}{3x^2}$

   (f) $y = 3x^{\frac{1}{2}} + x^{\frac{3}{2}} + x^{\frac{5}{2}} \Rightarrow \dfrac{dy}{dx} = \dfrac{3}{2}x^{-\frac{1}{2}} + \dfrac{3}{2}x^{\frac{1}{2}} + \dfrac{5}{2}x^{\frac{3}{2}} = \dfrac{3}{2\sqrt{x}} + \dfrac{3\sqrt{x}}{2} + \dfrac{5\sqrt{x^3}}{2}$

2. (a) $y = x^4 + 3x^3 \Rightarrow \dfrac{dy}{dx} = 4x^3 + 9x^2$ (b) $y = 2x^3 - 4x^2 + 3x - 6 \Rightarrow \dfrac{dy}{dx} = 6x^2 - 8x + 3$

   (c) $y = x^5 + 2x^3 + x \Rightarrow \dfrac{dy}{dx} = 5x^4 + 6x^2 + 1$ (d) $y = 1 + 3x^{-1} \Rightarrow \dfrac{dy}{dx} = -3x^{-2} = -\dfrac{3}{x^2}$

3. (a) $f(x) = 1 - 2x - 3x^2 \Rightarrow f'(x) = -2 - 6x$

   (b) $f(x) = x^4 + x^3 + x^2 + x \Rightarrow f'(x) = 4x^3 + 3x^2 + 2x + 1$

   (c) $f(x) = 24 + 3x - 8x^3 - x^4 \Rightarrow f'(x) = 3 - 24x^2 - 4x^3$

   (d) $f(x) = x^{\frac{5}{2}} + x^{\frac{3}{2}} + x^{-\frac{1}{2}} \Rightarrow f'(x) = \dfrac{5}{2}x^{\frac{3}{2}} + \dfrac{3}{2}x^{\frac{1}{2}} - \dfrac{1}{2}x^{-\frac{3}{2}} = \dfrac{5\sqrt{x^3}}{2} + \dfrac{3\sqrt{x}}{2} - \dfrac{1}{2\sqrt{x^3}}$

   (e) $f(x) = x^2 + 1 + \dfrac{1}{4}x^{-2} \Rightarrow f'(x) = 2x - \dfrac{1}{2}x^{-3} = 2x - \dfrac{1}{2x^3}$

4. (a) $\dfrac{dy}{dx} = 2x$, $x = 3$, so gradient is 6. (b) $f'(x) = -2x^{-3} - 12x^{-4} \Rightarrow f'(1) = -14$

   (c) $g'(x) = \dfrac{3}{2}x^{\frac{1}{2}} + x^{-2} \Rightarrow g'\left(\dfrac{1}{4}\right) = 3 + 16 = 19$ (d) $\dfrac{dy}{dx} = 3x^2 + 5$, $x = -2$, so gradient is 17.

5. (a) $\dfrac{dy}{dx} = 3x^2 = 48 \Rightarrow x = \pm 4$, so coordinates are $(-4, -64)$ and $(4, 64)$.

   (b) $f'(x) = 2x + 5 = 8 \Rightarrow x = \dfrac{3}{2}$, so coordinates are $\left(\dfrac{3}{2}, \dfrac{63}{4}\right)$.

   (c) $g'(x) = -4x^{-2} = -1 \Rightarrow x^2 = 4 \Rightarrow x = \pm 2$, so coordinates are $(-2, -1)$ and $(2, 3)$

6. $\dfrac{dy}{dx} = 2ax + b$. At $x = -2$, the gradient is -3, so $-4a + b = -3$. In addition, $-5 = 4a - 2b$.

   Therefore, $a = \dfrac{11}{4}$, $b = 8$.

7. $f(x) = 2x^2 - 7x + 3 \Rightarrow f'(x) = 4x - 7 \Rightarrow f'(5) = 13$. Therefore, the tangent is
   $y - 18 = 13(x - 5) \Rightarrow y = 13x - 47$ and $m = 13$, $c = -47$.

8. $\dfrac{dy}{dx} = 2x = 1$ at the tangent point. Therefore, at the tangent point, $x = \dfrac{1}{2}$ and $y = \dfrac{9}{2}$.

   Then, $\dfrac{9}{2} = \left(\dfrac{1}{2}\right)^2 + k \Rightarrow k = \dfrac{17}{4}$.

Exercise 7.5

1. (i) (a) $f(a) < 0$ (b) $f'(a) > 0$ (c) $f''(a) > 0$

   (ii) (a) $f(a) > 0$ (b) $f'(a) > 0$ (c) $f''(a) = 0$

   (iii) (a) $f(a) > 0$ (b) $f'(a) = 0$ (c) $f''(a) < 0$

   (iv) (a) $f(a) < 0$ (b) $f'(a) = 0$ (c) $f''(a) = 0$

2. (a) 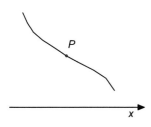   (b)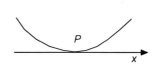

   (c) 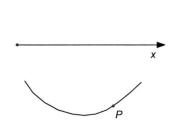   (d)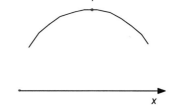

3. (a), (iv)   (b), (i)   (c), (ii)   (d), (iii)

4. (a)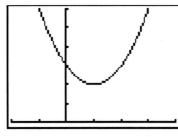
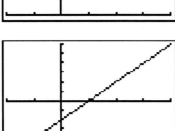

(b)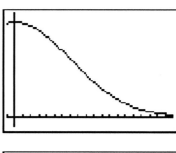
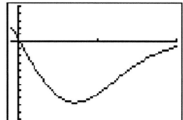

(c)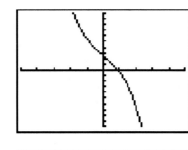

(d)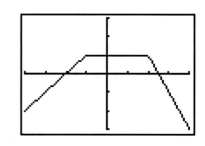
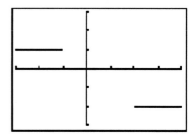

Exercise 7.6

1. (a) $f'(x) = -2 - 2x = 0 \Rightarrow x = -1$. So $f(-1) = 7$ is a maximum value

x	-2	-1	0
$f'(x)$	2	0	-2
Tangent	↗	→	↘

(b) $f'(x) = 2x + 7 = 0 \Rightarrow x = -\frac{7}{2}$. So $f(-3.5) = -\frac{65}{4}$ is a minimum value.

x	-4	-3.5	-3
$f'(x)$	-1	0	1
Tangent	↘	→	↗

(c) $f'(x) = 3x^2 - 10x - 8 = 0 \Rightarrow (3x+2)(x-4) = 0 \Rightarrow x = -\frac{2}{3}, x = 4$

x	-1	-2/3	0
$f'(x)$	5	0	-8
Tangent	↗	→	↘

x	3	4	5
$f'(x)$	-11	0	17
Tangent	↘	→	↗

So $f'\left(-\frac{2}{3}\right) = \frac{373}{27} \approx 13.8$ is a local maximum, and $f(4) \approx -37$ is a local minimum.

(d) $f'(x) = 6x^2 + 2x - 4 = 0 \Rightarrow 3x^2 + x - 2 = 0 \Rightarrow (x+1)(3x-2) = 0 \Rightarrow x = -1, x = \frac{2}{3}$

x	-2	-1	0
$f'(x)$	16	0	-4
Tangent	↗	→	↘

x	0	0.667	1
$f'(x)$	-4	0	4
Tangent	↘	→	↗

So $f(-1) = -2$ is a local maximum, and $f\left(\frac{2}{3}\right) = -\frac{179}{27} \approx -6.63$ is a local minimum.

(e) $f'(x) = 4x^3 - 256 = 0 \Rightarrow x^3 = 64 \Rightarrow x = 4$. So $f(4) = -753$ is a minimum.

x	3	4	5
$f'(x)$	-148	0	244
Tangent	↘	→	↗

(f) $f'(x) = \frac{1}{2\sqrt{x}} - 2 = 0 \Rightarrow 4\sqrt{x} = 1 \Rightarrow x = \frac{1}{16}$. So $f\left(\frac{1}{16}\right) = \frac{1}{8}$ is a maximum.

x	1/36	1/16	1
$f'(x)$	1	0	-1.5
Tangent	↗	→	↘

2. (a) (i) $\frac{dy}{dx} = 2x + 6$

   (ii) $2x + 6 = 0 \Rightarrow x = -3$

   (iii) $(-3, -1)$, $f''(x) = 2 > 0$ so $(-3, -1)$ has a minimum value.

(b) (i) $\dfrac{dy}{dx} = 3x^2 - 12$

(ii) $3x^2 - 12 = 0 \Rightarrow x = \pm 2$

(iii) $(-2, 19)$, $(2, -13)$, $f''(x) = 6x \Rightarrow f''(-2) = -12 < 0$ and $f''(2) = 12 > 0$. So $(-2, 19)$ is a local maximum and $(2, -13)$ is a local minimum.

(c) (i) $\dfrac{dy}{dx} = 5x^4 - 80$

(ii) $5x^4 - 80 = 0 \Rightarrow x = \pm 2$

(iii) $(-2, 128)$, $(2, -128)$, $f''(x) = 20x^3 \Rightarrow f''(-2) = -160 < 0$ and $f''(2) = 160 > 0$. So $(-2, 128)$ is a local maximum and $(2, -128)$ is a local minimum.

3. (a) (i) $f'(x) = 3x^2 + 9 \Rightarrow f''(x) = 6x$

(ii) At points of inflection $f''(x) = 0$ so that $6x = 0 \Rightarrow x = 0$. Then $f(0) = -7$ and the coordinates of the point of inflection are $(0, -7)$.

(b) (i) $f'(x) = 2x + x^{-2} \Rightarrow f''(x) = 2 - 2x^{-3} = 2\left(1 - \dfrac{1}{x^3}\right)$.

(ii) At points of inflection $f''(x) = 0$ so that $1 - \dfrac{1}{x^3} = 0 \Rightarrow x^3 = 1 \Rightarrow x = 1$. Then $f(1) = 0$ and the coordinates of the point of inflection are $(1, 0)$.

(c) (i) $f'(x) = 4x^3 + 9x^2 + 6x - 1 \Rightarrow f''(x) = 12x^2 + 18x + 6$

(ii) At points of inflection $f''(x) = 0$ so that $2x^2 + 3x + 1 = 0 \Rightarrow (2x+1)(x+1) = 0$
$\Rightarrow x = -1,\ x = -\dfrac{1}{2}$. Then $f(-1) = 7$ and $f\left(-\dfrac{1}{2}\right) = \dfrac{95}{16} \approx 5.94$ and the coordinates of the two points of inflection are $(-1, 7)$ and $(-0.5, 5.94)$.

Exercise 7.7

1. (a) $y = \dfrac{8-x}{3}$ (b) $W = \dfrac{8x}{3} - \dfrac{x^2}{3}$ (c) $\dfrac{dW}{dx} = \dfrac{8}{3} - \dfrac{2x}{3}$ (d) $\dfrac{8}{3} - \dfrac{2x}{3} = 0 \Rightarrow x = 4$

   (e)

x	3	4	5
$\dfrac{dW}{dx}$	$\dfrac{2}{3}$	0	$-\dfrac{2}{3}$
Tangent	↗	→	↘

   Therefore W has a maximum value of $\dfrac{16}{3}$ when $x = 4$.

2. (a) $P(x) = (400 - 2x)x - (10000 - 400x + 8x^2) = 800x - 10000 - 10x^2$

   (b) $P(x) = 800 - 20x = 0 \Rightarrow x = 40$. Therefore, as $400 - 2 \times 40 = 320$, the profit is maximized when each digital camera is sold for \$320.

   (c) $P''(x) = -20 < 0 \Rightarrow P$ has a maximum value when $x = 40$.

3. (a) Area $A = y^2$, but $NP = \sqrt{y^2 - x^2}$ and as $x + \sqrt{y^2 - x^2} = 8$, $\sqrt{y^2 - x^2} = 8 - x$
   $\Rightarrow y^2 - x^2 = (8-x)^2 = 64 - 16x + x^2 \Rightarrow A = y^2 = 64 - 16x + 2x^2$.

   (b) $A'(x) = -16 + 4x$, $A'(x) = 0 \Rightarrow -16 + 4x = 0 \Rightarrow x = 4$. So A is minimized when $x = 4$.

x	3	4	5
$A'(x)$	-4	0	4
Tangent	↘	→	↗

4. (a) Each side of the square is x. There are 4 sides, so the perimeter of the square cross-section is $4x$. Hence the total perimeter is $4x + l$. Therefore, $4x + l = 6$.

   (b) $V = lx^2 = x^2(6 - 4x) = 6x^2 - 4x^3$

   (c) $V'(x) = 12x - 12x^2$, $V'(x) = 0 \Rightarrow 12x - 12x^2 = 0 \Rightarrow 12x(1-x) = 0 \Rightarrow x = 0, 1$

   (d) $V''(x) = 12 - 24x \Rightarrow V''(1) = 12 - 24 < 0$. When $x = 1$, the volume has a maximum value. This value is $V(1) = 2$ and the maximum volume is 2m^2.

5. (a) $2x^2h = 50 \Rightarrow h = \dfrac{25}{x^2}$

   (b) Surface area, $S = 2xh + 4xh + 2x^2 = 6xh + 2x^2 = 6x\left(\dfrac{25}{x^2}\right) + 2x^2 = \dfrac{150}{x} + 2x^2$

   $S'(x) = 4x - \dfrac{150}{x^2} = 0 \Rightarrow 4x^3 = 150 \Rightarrow x \approx 3.34716$. Then $h \approx 2.23145$ and dimensions of the box that minimize the surface area are, in centimeters, $3.35 \times 6.69 \times 2.23$.

   (c) $S''(x) = 4 + \dfrac{300}{x^3} \Rightarrow S''(3.34716) = 12 > 0$, and so the dimensions give a minimum rather than a maximum surface area.

6. (a) $\pi x^2 h = 500 \Rightarrow h = \dfrac{500}{\pi x^2}$

   (b) $A = 2\pi x^2 + 2\pi xh = 2\pi x^2 + 2\pi x\left(\dfrac{500}{\pi x^2}\right) = 2\pi x^2 + \dfrac{1000}{x}$

   (c) $A'(x) = 4\pi x - \dfrac{1000}{x^2} = 0 \Rightarrow x^3 = \dfrac{250}{\pi} \Rightarrow x = 4.3013$, $A(4.3013) \approx 348.7$

   $A''(x) = 4\pi + \dfrac{2000}{x^3} \Rightarrow A''(4.3013) = 37.7 > 0$, ensuring that $349\,\text{cm}^2$ is the minimum, as opposed to the maximum surface area.

7. (a) Area, $A = PQ \times PS$, but if $OP = x$ and $PS = y$,
   $PQ + 2x = 1$ and $PS = x - x^2$.
   $\Rightarrow A = (1 - 2x)(x - x^2) = 2x^3 - 3x^2 + x$

   (b) $A'(x) = 6x^2 - 6x + 1$ and $A'(x) = 0 \Rightarrow 6x^2 - 6x + 1 = 0$.

   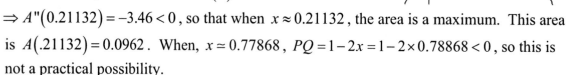

   $\Rightarrow \left(x - \dfrac{1}{2}\right)^2 = \dfrac{1}{12} \Rightarrow x - \dfrac{1}{2} = \pm\dfrac{1}{2\sqrt{3}}$

   $\Rightarrow x = 0.21132$, $x = 0.78868$. $A''(x) = 12x - 6$

   $\Rightarrow A''(0.21132) = -3.46 < 0$, so that when $x \approx 0.21132$, the area is a maximum. This area is $A(.21132) = 0.0962$. When, $x \approx 0.77868$, $PQ = 1 - 2x = 1 - 2 \times 0.78868 < 0$, so this is not a practical possibility.

Exercise 7.8

1. (a) $f'(x) = 9(3x+1)^2$ (b) $f'(x) = -16x(1-4x^2)$

   (c) $f(x) = (x^2+5)^{\frac{1}{2}} \Rightarrow f'(x) = \frac{1}{2}(x^2+5)^{-\frac{1}{2}} \times 2x = \frac{x}{\sqrt{x^2+5}}$

   (d) $g'(x) = -e^{-x}$ (e) $g'(x) = 12e^{4x}$ (f) $g'(x) = -2xe^{1-x^2}$ (g) $h'(x) = 3\cos 3x$

   (h) $h'(x) = -2\sin\left(2x - \frac{\pi}{6}\right)$ (i) $h'(x) = 2\sin x \cos x$ (j) $f'(x) = \frac{1}{1+5x} \times 5 = \frac{5}{1+5x}$

   (k) $f'(x) = \frac{4}{x^2+3} \times 2x = \frac{8x}{x^2+3}$ (l) $f'(x) = \frac{1}{1+e^x} \times e^x = \frac{e^x}{1+e^x}$

2. (a) $\frac{dy}{dx} = 4(2x-5)$ (b) $\frac{dy}{dx} = 6x(x^2-4)^2$ (c) $\frac{dy}{dx} = -20(3-5x)^3$ (d) $\frac{dy}{dx} = 2xe^{x^2}$

   (e) $\frac{dy}{dx} = -2e^{1-2x}$ (f) $\frac{dy}{dx} = -3\cos^2 x \sin x$ (g) $\frac{dy}{dx} = 12\cos 4x$ (h) $\frac{dy}{dx} = 6x\cos 3x^2$

   (i) $y = \ln\frac{1}{x} = \ln x^{-1} = -\ln x \Rightarrow \frac{dy}{dx} = -\frac{1}{x}$ (j) $\frac{dy}{dx} = \frac{3x^2}{(x^3-3)}$ (k) $\frac{dy}{dx} = \frac{4}{4x} = \frac{1}{x}$

   (l) $\frac{dy}{dx} = -2(\cos x)^{-2}(-\sin x) = \frac{2\sin x}{\cos^2 x}$

3. (a) $f'(x) = -28(1-7x)^3$ (b) $f'(x) = -20e^{-4x}$ (c) $\frac{dy}{dx} = \frac{1+\cos x}{x+\sin x}$ (d) $\frac{dy}{dx} = \frac{-2x}{(x^2-3)^2}$

   (e) $g'(x) = 2\sin x(2-\cos x)$ (f) $f'(x) = e^{-x}\sin(e^{-x})$ (g) $f'(x) = -6\cos\left(\frac{\pi}{2} - 6x\right)$

   (h) $\frac{dy}{dx} = 3e^x(1+e^x)^2$ (i) $\frac{dy}{dx} = \frac{\cos x}{2\sqrt{\sin x + 1}}$ (j) $f'(x) = \sin x e^{-\cos x}$ (k) $\frac{dy}{dx} = \frac{-2\sin 2x}{\cos 2x}$

   (l) $g'(x) = 6\sin 3x \cos 3x$

Exercise 7.9

1. (a) $f'(x) = x^2 \cos x + 2x \sin x$

   (b) $f'(x) = x^3 2(x+1) + 3x^2(x+1)^2 = x^2(x+1)(2x+3(x+1)) = x^2(x+1)(5x+3)$

   (c) $f'(x) = e^x(-\sin x) + e^x \cos x = e^x(\cos x - \sin x)$

   (d) $f'(x) = x\dfrac{1}{x} + 1\ln x = 1 + \ln x$

   (e) $f'(x) = 4xe^x + 4e^x = 4e^x(x+1)$

   (f) $f'(x) = 1(\cos x + \sin x) + x(-\sin x + \cos x) = (1-x)\sin x + (1+x)\cos x$

2. (a) $f'(x) = \dfrac{2x}{(x^2+1)^2}$ (b) $f'(x) = \dfrac{\ln x - 1}{(\ln x)^2}$ (c) $f'(x) = \dfrac{x \cos x - \sin x}{x^2}$

   (d) $f'(x) = \dfrac{-\ln x}{x^2}$ (e) $f'(x) = \dfrac{1}{\cos^2 x}$ (f) $f'(x) = \dfrac{-1 - 2\sin x}{(2 + \sin x)^2}$

3. (a) $\dfrac{dy}{dx} = \dfrac{(2x+1)1 - 2x}{(2x+1)^2} = \dfrac{1}{(2x+1)^2}$

   (b) $\dfrac{dy}{dx} = x\left(-2(2x+1)^{-2}\right) + 1(2x+1)^{-1} = \dfrac{-2x + (2x+1)}{(2x+1)^2} = \dfrac{1}{(2x+1)^2}$

4. (a) The denominator is a constant, so differentiate directly to obtain $\dfrac{dy}{dx} = \dfrac{1}{3}(1 + \cos x)$.

   (b) (i) $\dfrac{dy}{dx} = \dfrac{(x \sin x)0 - (1 + \cos x)3}{(x + \sin x)^2} = \dfrac{-3(1 + \cos x)}{(x + \sin x)^2}$

   (ii) $\dfrac{dy}{dx} = 3\left(-1(x + \sin x)^{-2}\right)(1 + \cos x) = \dfrac{-3(1 + \cos x)}{(x + \sin x)^2}$

5. (a) $f'(x) = 2x \cos 2x - x^2 2 \sin 2x = 2x(\cos 2x - x \sin 2x)$ (b) $g'(x) = \dfrac{2 - 2x^2}{(x^2 + 1)^2}$

(c) $\dfrac{dy}{dx} = \dfrac{1-\ln x}{x^2}$ (d) $\dfrac{dy}{dx} = x3(1-4x)^2(-4) + (1-4x)^3 = (1-4x)^2(1-16x)$

(e) $g'(x) = x+2$ (f) $f'(x) = -e^{-x}(x+2)$ (g) $\dfrac{dy}{dx} = \dfrac{1}{(2x+3)^2}$

(h) $g(x) = \ln(x^2-1) - \ln(x^2+1) \Rightarrow g'(x) = \dfrac{2x}{x^2-1} - \dfrac{2x}{x^2+1} = \dfrac{4x}{x^4-1}$ (i) $\dfrac{dy}{dx} = \dfrac{e^{2x}(5-6x)}{(1-3x)^2}$

(j) $f'(x) = e^{-x}(2\cos 2x - \sin 2x)$ (k) $\dfrac{dy}{dx} = \dfrac{6}{\cos^2 3x}$

(l) $g'(x) = 2\cos 2x \cos x - \sin 2x \sin x$

6. (a) $\dfrac{dy}{dx} = \dfrac{(1+e^x)e^x - e^x \times e^x}{(1+e^x)^2} = \dfrac{e^x + e^{2x} - e^{2x}}{(1+e^x)^2} = \dfrac{e^x}{(1+e^x)^2}$

(b) $f'(x) = \dfrac{(\sin x + \cos x)(\cos x + \sin x) - (\sin x - \cos x)(\cos x - \sin x)}{(\sin x + \cos x)^2}$

$= \dfrac{1 + 2\sin x \cos x + 1 - 2\sin x \cos x}{\sin^2 x + 2\sin x \cos x + \cos^2 x} = \dfrac{2}{1+\sin 2x}$

(c) $g(x) = \ln x - \ln(2x+1) \Rightarrow g'(x) = \dfrac{1}{x} - \dfrac{2}{2x+1} = \dfrac{2x+1-2x}{x(2x+1)} = \dfrac{1}{x(2x+1)}$

Exercise 7.10

1. (a) $f'(x) = -2 < 0$ for all $x \in \mathbb{R}$, so $f(x)$ is a decreasing function.

(b) $f'(x) = -\dfrac{3}{x^4}$ but $x^4 > 0$ for all x so $-\dfrac{3}{x^4} < 0$ for all x, so $f(x)$ is a decreasing function.

(c) $f'(x) = \dfrac{1}{x} > 0$ for $x > 0$, so $f(x)$ is an increasing function.

(d) $f'(x) = -e^{-x}$. As $e^{-x} > 0$ for all $x \in \mathbb{R}$, $-e^{-x} < 0$ for all x, so $f(x)$ is a decreasing function.

(e) $f'(x) = 3x^2 + 1$, but $x^2 \geq 0 \Rightarrow 3x^2 + 1 > 0$, so $f(x)$ is an increasing function.

2. (a) $\dfrac{dy}{dx} = e^{-x}(1-x)$. For $x > 1$, $1-x < 0$, $e^{-x} > 0 \Rightarrow \dfrac{dy}{dx} < 0$. Therefore, $y = xe^{-x}$ is a decreasing function for $x > 1$.

(b) $\dfrac{dy}{dx} = k - \cos x$. For $y = kx - \sin x$ to be an increasing function, $k - \cos x > 0$ for all x. But $-1 \leq \cos x \leq 1$, therefore, if $k > 1$, $k - \cos x > 0$ for all x. Therefore, $k > 1$.

3. (a) $12x - y - 16 = 0$ (b) $3x - y - 2 = 0$ (c) $2x - y + 4 = 0$ (d) $4x + y + 32 = 0$

(e) $x - 4y + 4 = 0$

4. (a) $x + 12y - 98 = 0$ (b) $x + 3y - 4 = 0$ (c) $x + 2y + 2 = 0$ (d) $x - 4y - 196 = 0$

(e) $4x + y - 18 = 0$

5. $y = x$

6. $\dfrac{dy}{dx} = 2x + 2$. At $x = -2$ the gradient of the tangent is -2, so the gradient of the tangent is $\dfrac{1}{2}$.

When $x = -2$, $y = 5$, the equation of the normal is $y - 5 = \dfrac{1}{2}(x+2) \Rightarrow 2y - 10 = x + 2$

$\Rightarrow x - 12y + 12 = 0$.

7. $\dfrac{dy}{dx} = x(1 + 2\ln x)$, so when $x = e$, the gradient is $e(1+2) = 3e$ and $y = e^2$. Therefore, the equation of the tangent is $y - e^2 = 3e(x-e) \Rightarrow y = 3ex - 2e^2$.

8. $f'(x) = xe^{-x}(2-x) \Rightarrow f'(1) = \dfrac{1}{e}$. $f(1) = \dfrac{1}{e}$. Therefore, the equation of the tangent is

$y - \dfrac{1}{e} = \dfrac{1}{e}(x-1) \Rightarrow y = \dfrac{x}{e}$.

9. $f'(x) = \dfrac{e^{2x}(2x-1)}{x^2} \Rightarrow f'(1) = e^2$. $f(x) = e^2$, so the equation of the tangent is

$y - e^2 = e^2(x-1) \Rightarrow y = e^2 x$.

10. $f'(x) = 1 + \ln x \Rightarrow f'(e) = 2$, $f(e) = e$. The gradient of the normal is $-\dfrac{1}{2}$. Therefore, the equation of the normal is $y - e = -\dfrac{1}{2}(x-e) \Rightarrow 2y - 2e = -x + e \Rightarrow x + 2y - 3e = 0$.

Exercise 7.11

1. (a) $f'(x) = \dfrac{2(4-x^2)}{(x^2+4)^2}$, $f'(x) = 0 \Rightarrow 2(4-x^2) = 0 \Rightarrow x^2 = 4 \Rightarrow x = \pm 2$

   (b)
x	1	2	3
$f'(x)$	6/25	0	-10/169
Tangent	↗	→	↘

   (c) Minimum value is $f(-2) = -\dfrac{1}{2}$.

2. (a) $f'(x) = e^{-x}(1-x)$

   (b) $f'(x) = 0 \Rightarrow e^{-x}(1-x) = 0 \Rightarrow x = 1$. Therefore, a maximum or minimum occurs when $x = 1$.

   (c) $f''(x) = e^{-x}(x-2) \Rightarrow f''(1) = -e^{-1} < 0$, so $f(1)$ is a maximum.

   (d) Maximum value is $f(1) = \dfrac{1}{e}$.

3. (a)

   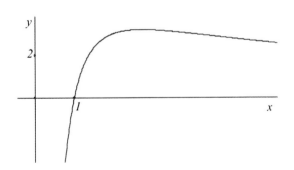

   (b) $\dfrac{dy}{dx} = \dfrac{1-\ln x}{x^2}$, $\dfrac{d^2y}{dx^2} = \dfrac{x^2(-1/x) - (1-\ln x)2x}{x^4} = \dfrac{-x - 2x + 2x\ln x}{x^4} = \dfrac{2\ln x - 3}{x^3}$

   (c) $\dfrac{dy}{dx} = 0 \Rightarrow 1 - \ln x = 0 \Rightarrow \ln x = 1 \Rightarrow x = e$. When $x = e$, $\dfrac{d^2y}{dx^2} = -\dfrac{1}{e^3} < 0$ so y has a maximum value.

   (d) $y = \dfrac{1}{e}$

4. (a) $\dfrac{dy}{dx} = e^{-x}\left(-\dfrac{1}{2}\sin\dfrac{1}{2}x\right) + \left(e^{-x}\right)\cos\dfrac{1}{2}x = -\dfrac{1}{2}e^{-x}\left(\sin\dfrac{1}{2}x + 2\cos\dfrac{1}{2}x\right)$

(b) $\sin\dfrac{1}{2}x + 2\cos\dfrac{1}{2}x = 0 \Rightarrow \tan\dfrac{1}{2}x = -2 \Rightarrow \dfrac{1}{2}x = \pi - 1.1071,\ 2\pi - 1.1071 \Rightarrow x \approx 4.069,\ 10.352$

but as $10.352 > 2\pi$, the only solution in the interval is $x = 4.07$.

(c)

x	3	4.07	5
$\dfrac{dy}{dx}$	-.028	0	.0034
Tangent	↘	→	↗

5. (a)

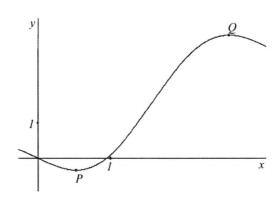

(b) $g'(x) = 1 - 2\cos 2x$

$g'(x) = 0 \Rightarrow 1 - 2\cos 2x = 0 \Rightarrow \cos 2x = \dfrac{1}{2} \Rightarrow 2x = \dfrac{\pi}{3}, \dfrac{5\pi}{3} \Rightarrow x = \dfrac{\pi}{6}, \dfrac{5\pi}{6}$

(c) $g''(x) = 4\sin 2x \Rightarrow g''\left(\dfrac{\pi}{6}\right) = 4\sin\dfrac{\pi}{3} = 2\sqrt{3} > 0$. So value of function at P is

$f\left(\dfrac{\pi}{6}\right) = \dfrac{\pi}{6} - \sin\dfrac{\pi}{3} = \dfrac{\pi}{6} - \dfrac{\sqrt{3}}{2} = \dfrac{\pi - 3\sqrt{3}}{6}$ and the coordinates of P are $\left(\dfrac{\pi}{6}, \dfrac{\pi - 3\sqrt{3}}{6}\right)$.

(d) Q is $\left(\dfrac{5\pi}{6}, \dfrac{5\pi + 3\sqrt{3}}{6}\right)$, and is a maximum value, because $f''\left(\dfrac{5\pi}{6}\right) = 4\sin\dfrac{5\pi}{3} = -2\sqrt{3} < 0$.

6. (a) $\pi r + 2x = 5$

(b) $A = \pi r^2 + x^2$, $r = \dfrac{5 - 2x}{\pi} \Rightarrow A = \pi\left(\dfrac{5 - 2x}{\pi}\right)^2 + x^2 = x^2 + \dfrac{(5 - 2x)^2}{\pi}$

(c) $A'(x) = 2x - \dfrac{4}{\pi}(5-2x) = 0 \Rightarrow 2\pi x - 20 + 8x = 0 \Rightarrow x = \dfrac{10}{\pi+4} \approx 1.40$. $A''(x) = 2 + \dfrac{8}{\pi} > 0$, so $A(1.40)$ has a minimum value.

7. (a) $L = \sqrt{(x-0)^2 + (9.5-x^2)^2} = \sqrt{x^2 + 9.5^2 - 19x^2 + x^4} = \sqrt{x^4 - 18x^2 + 90.25}$

   (b) $\dfrac{dL}{dx} = \dfrac{2x(x^2-9)}{\sqrt{x^4 - 18x^2 + 90.25}}$

   (c) $\dfrac{dL}{dx} = 0 \Rightarrow 2x(x^2-9) = 0 \Rightarrow x = 0, x = 3$ and $x = -3$

   (d) Points closest to P are $(3, 9), (-3, 9)$.

   (e) O is a local maximum. The origin, O is further from P than all other points close to O.

8. (a) $P(x) = 0.85 f(x) - (8x+30) = 425(1 - e^{-0.2x}) - 8x - 30$

   (b) $P'(x) = 85e^{-0.2x} - 8$, $P'(x) = 0 \Rightarrow 85e^{0.2x} = 8 \Rightarrow x \approx 11.8$, but as $x \in \mathbb{Z}^+$, $x = 12$. Therefore, 12 employees gives maximum profit of $P(12) = 260.44$.

   (c)

   ```
   Maximum
   X=11.816046  Y=260.47161
   ```

9. (a) $P(x) = 8x - \left(3x + \dfrac{x(x-400)^2}{80000}\right) = 5x - \dfrac{x(x-400)^2}{80000}$

   (b) $P(300) = 5 \times 300 - \dfrac{300(-100)^2}{80000} 1500 - \dfrac{300}{8} = 1500 - 37.5 = 1462.50$

   (c) $\dfrac{dP}{dx} = 5 - \dfrac{1}{80000}\left(x \cdot 2(x-400) + (x-400)^2\right) = 5 - \dfrac{(x-400)(3x-400)}{80000}$

$$\frac{dP}{dx} = 0 \Rightarrow 5 - \frac{(x-400)(3x-400)}{80000} = 0 \Rightarrow (x-400)(3x-400) = 400\ 000$$

$\Rightarrow 3x^2 - 1600x - 240000 = 0 \Rightarrow x = -122.1,\ 655.4$. But $x > 0$, so that the maximum profit occurs when 655 liters of wine is produced. This profit is $\$P(655.4) = \$2742.61$.

(d)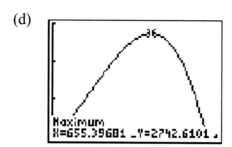

10. (a) For $x \le 40$, $P(x) = 0.9xM(x) - 0.65x = 0.9 \times 8 \times x - 0.65x = 6.55x$. For $x > 40$,

$$P(x) = 0.9xM(x) - 0.65x = 0.9x\left(8 - \frac{1}{200}(x-40)^2\right) - 0.65x.$$

(b) $P(x) = 6.55x - \dfrac{0.9x}{200}(x-40)^2$. $P'(x) = 6.55 - \dfrac{0.9}{200}(x-40)(3x-40) = 0$ for maximum profit $\Rightarrow 3x^2 - 160x + 144.44 = 0 \Rightarrow x = 52.4,\ 0.91$. But $x \in \mathbb{Z}^+$, so the number of cows that maximizes profit is 52.

(c) $P(30) = 6.55 \times 30 = 196.50$. Therefore, Jane makes a daily profit of $196.50.

(d) Now the situation changes because each pays for their own costs, but both suffer the consequences of there being too many cows in the field. Therefore, Kathy's profit is

$$P(30) = 0.9 \times 30\left(8 - \frac{1}{200}(60-40)^2\right) - 0.65 \times 30 = 142.50,$$ so she makes $142.50 per day.

(e) Jane's profit has been reduced to $142.50.

Exercise 7.12

1. (a) 5 m

(b) $v(t) = y'(t) = 12 - 2t \Rightarrow v(0) = 12$. So the initial velocity is $12\ \text{ms}^{-1}$.

(c) Maximum displacement occurs when $v(t) = 0 \Rightarrow t = 6$. So the maximum displacement is $y(6) = 41\ \text{m}$.

(d) $12 - 2t = 10 \Rightarrow t = 1$

2. (a) $x'(t) = 3t^2 - 10t + 4 = 0 \Rightarrow t = 0.465, 2.87$

   (b) $t^3 - 5t^2 + 4t = 0 \Rightarrow t(t-1)(t-4) = 0 \Rightarrow t = 0, 1, 4$. Therefore, the particle returns after 1 second and 4 seconds.

   (c) The acceleration function is $a(t) = 6t - 10$. Therefore, (i) $a(2) = 2$ (ii) $a(5) = 20$

3. (a) $a(t) = v'(t) = 15\cos 3t$

   (b) $5\sin 3t = 0 \Rightarrow \sin 3t = 0 \Rightarrow 3t = \pi \Rightarrow t = \dfrac{\pi}{3}$. Therefore, the particle first comes to instantaneous rest after $\dfrac{\pi}{3} \approx 1.05$ seconds.

   (c) Maximum acceleration is $15\,\text{ms}^{-2}$.

4. (a) $s(5.2) = 1.4 - 5(5.2)^2 = -133.8$. So the estimated depth of the mine shaft is 133.8m or about 134m.

   (b) $v(t) = -10t$ so $v(5.2) = -52$ and the speed of the stone is $52\,\text{ms}^{-1}$.

5. (a) $v(t) = x'(t) = \dfrac{4t}{1+2t^2}$

   (b) $v'(t) = \dfrac{4(1-2t^2)}{(1+2t^2)^2}$. Maximum velocity occurs when $v'(t) = 0 \Rightarrow 1-2t^2 = 0 \Rightarrow t = \pm\dfrac{1}{\sqrt{2}}$.

   Therefore, $v\left(\pm\dfrac{1}{\sqrt{2}}\right) = \dfrac{4(\pm 1/\sqrt{2})}{(1+2(1/2))} = \pm\dfrac{4}{2\sqrt{2}} = \pm\sqrt{2}$ and the maximum velocity is $\sqrt{2}\,\text{ms}^{-1}$.

   (c) Average velocity $= \dfrac{\text{distance}}{\text{time}} = \dfrac{\ln 19 - \ln 3}{3-1} \approx 0.923\,\text{ms}^{-1}$.

   (d) The maximum acceleration is $4\,\text{ms}^{-2}$ which occurs when $t = 0$.

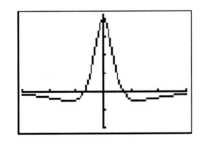

6. (a) $v(t) = x'(t) = \cos t e^{\sin t}$

(b)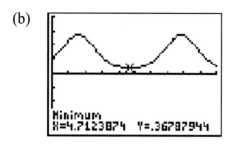

So minimum displacement from O is 0.368 meters occurring after 4.71 seconds.

(c) The maximum displacement occurs when $v(t) = 0 \Rightarrow \cos t e^{\sin t} = 0 \Rightarrow \cos t = 0 \Rightarrow t = \dfrac{\pi}{2}$.

The maximum displacement is $e^{\sin \frac{\pi}{2}} = e^1 = e$.

(d) The large sketch is not drawn but the shape of the velocity function can be seen from the graphing calculator screen display shown below.

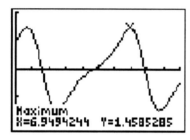

So the maximum velocity is 1.45 ms^{-1}.

Exercise 7.13

1. (a) $x^3 + c$
   (b) $\dfrac{1}{5}x^5 + c$
   (c) $-x^{-1} + c$
   (d) $\dfrac{1}{3}x^3 + \dfrac{1}{2}x^2 + x + c$

   (e) $2x - \dfrac{5}{2}x^2 + c$
   (f) $x^3 + x^2 + c$
   (g) $\dfrac{3}{4}x^{\frac{4}{3}} + c$
   (h) $2\sqrt{x} + c$

   (i) $3 \ln x + c$
   (j) $\dfrac{1}{2}x^4 - x^3 + c$
   (k) $\dfrac{1}{4}x^4 + \dfrac{3}{2}x^2 - \ln x + c$
   (l) $\dfrac{2}{3}\sqrt{x^3} + 2\sqrt{x} + c$

2. (a) $-3\cos x + c$
   (b) $\dfrac{1}{5}\sin x + c$
   (c) $\sin x - \cos x + x + c$
   (d) $4e^x + c$

   (e) $e^x - \sin x + c$
   (f) $2\sin x + \cos x + e^x + c$

3. (a) $\int \sin\frac{1}{2}x\cos\frac{1}{2}x\,dx = \int \frac{1}{2}\sin x\,dx = -\frac{1}{2}\cos x + c$

(b) $\int \sin^2\frac{1}{2}x\,dx = \frac{1}{2}\int(1-\cos x)\,dx = \frac{1}{2}(x-\sin x)+c = \frac{1}{2}x - \frac{1}{2}\sin x + c$

(c) $\int \cos^2\frac{1}{2}x\,dx = \frac{1}{2}\int(1+\cos x)\,dx = \frac{1}{2}(x+\sin x)+c = \frac{1}{2}x + \frac{1}{2}\sin x + c$

4. (a) $\frac{1}{x} - \frac{1}{x+1} = \frac{(x+1)-x}{x(x+1)} = \frac{1}{x(x+1)}$

(b) $\int \frac{1}{x(x+1)}dx = \int \frac{1}{x}dx - \int \frac{1}{x+1}dx = \ln x - \ln(x+1) + c = \ln\left(\frac{x}{x+1}\right) + c$

Exercise 7.14

1. $\frac{1}{8}(2x+1)^4 + c$
2. $-\frac{1}{3}(1-x)^3 + c$
3. $-\frac{1}{2}\cos 2x + c$
4. $\frac{1}{4}\sin 2x + c$

5. $\frac{1}{3}\ln(3x-2) + c$
6. $e^{x+2} + c$
7. $-\frac{1}{5}e^{3-5x} + c$
8. $-20e^{-0.05x} + c$

9. $-4\ln(10-x) + c$
10. $\frac{1}{3}\sin\left(3x - \frac{\pi}{4}\right) + c$
11. $\frac{1}{2}\cos\left(\frac{\pi}{3} - 2x\right) + c$

12. $\frac{1}{6}\sqrt{(4x+3)^3} + c$
13. $3\ln(x+2) + c$
14. $-\frac{1}{20}(3-4x)^5 + c$
15. $2\sqrt{2x+1} + c$

16. $\frac{1}{3}e^{3x} + c$
17. $\frac{1}{10}(5x-2)^2 + c$
18. $5\sin\frac{1}{5}x + \frac{5}{2}\cos\frac{2}{5}x + c$

19. $\frac{3}{\pi}\sin\frac{\pi x}{3} + c$
20. $-\frac{180}{\pi}\cos x° + c$

Exercise 7.15

1. (a) $y = 2x^2 + x + c$, $3 = 0 + 0 + c \Rightarrow c = 3 \Rightarrow y = 2x^2 + x + 3$

(b) $y = 3x - \frac{2}{3}x^3 + c$, $15 = 9 - 18 + c \Rightarrow c = 24 \Rightarrow y = 24 + 3x - \frac{2}{3}x^3$

(c) $y = -\cos x + c$, $1 = -\cos\frac{\pi}{3} + c \Rightarrow c = 1 + \cos\frac{\pi}{3} = \frac{3}{2} \Rightarrow y = \frac{3}{2} - \cos x$

(d) $y = -2e^{-2x} + c$, $1 = -2 + c \Rightarrow c = 3 \Rightarrow y = 3 - 2e^{-2x}$

(e) $y = (2x+1)^{\frac{1}{2}} + c$, $4 = 3 + c \Rightarrow c = 1 \Rightarrow y = \sqrt{2x+1} + 1$

2. $y = x^4 - 3x^2 + 5x + c$, $8 = 81 - 27 + 15 + c \Rightarrow c = -61 \Rightarrow y = x^4 - 3x^2 + 5x - 61$

3. $x = t^3 + \frac{2}{t} + c$, $5 = 1 + 2 + c \Rightarrow c = 2 \Rightarrow x = t^3 + \frac{2}{t} + 2$

4. $y = e^x + c$, $2 = 1 + c \Rightarrow c = 1 \Rightarrow y = e^x + 1$

5. $y = 4\ln x + c$, $5 = 4\ln e^2 + c \Rightarrow 5 = 8 + c \Rightarrow c = -3 \Rightarrow y = 4\ln x - 3$

Exercise 7.16

1. (a) $a = 10 \Rightarrow v(t) = -10t + c$, $v(0) = 16 \Rightarrow v(t) = 16 - 10t$

    (b) $v(2) = 16 - 20 = -4$. Therefore the velocity is -4ms^{-1}.

    (c) $s(t) = 16t - 5t^2 + c$, $s(0) = 2 \Rightarrow 2 = 0 + 0 + c \Rightarrow c = 2$. The maximum velocity occurs when $v(t) = 0 \Rightarrow t = 1.6 \Rightarrow s(1.6) = 16 \times 1.6 - 5(1.6)^2 + 2 = 14.8$.

2. (a) $a(5) = 5$

    (b) $v(t) = 20t - \frac{3}{2}t^2 + c$, $3 = 0 + 0 + c \Rightarrow c = 3 \Rightarrow v(t) = 20t - \frac{3}{2}t^2 + 3$

    (c) $v(5) = 65.5$

    (d) $20t - \frac{3}{2}t^2 + 3 = 0 \Rightarrow 3t^2 - 40t - 6 = 0 \Rightarrow t \approx -0.148, 13.482$. But $t > 0$, so $t = 13.5$.

3. (a) $v(10) = 306.8$, so the velocity of the rocket is 307ms^{-1}

    (b) $s(t) = \frac{1}{2}t^2 + 4e^{0.5t} + c$, $s(0) = 0 + 4 + c = 0 \Rightarrow c = -4 \Rightarrow s(t) = \frac{1}{2}t^2 + 4e^{0.5t} - 4$

(c) $s(10) = 639.7$, therefore the altitude is 640m.

(d) For a further 50s the rocket travels at 306.8ms^{-1}. Therefore after 1 minute, its height is $639.7 + 306.8 \times 50 = 15976.7$ or approximately 16 000m.

4. (a) (i) $v(t) = -4\cos t + c$, $8 = -4 + c \Rightarrow c = 12$ and $v(t) = 12 - 4\cos t$.

(ii) Maximum velocity is 16ms^{-1} occurring when $\cos t = -1 \Rightarrow t = \pi \approx 3.14s$.

(iii) $s(t) = 12t - 4\sin t + c$, $c = 0 \Rightarrow s(t) = 12t - 4\sin t$

(b)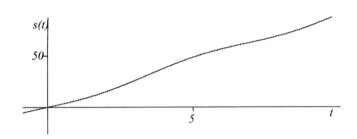

5. $v(0) = 10 \Rightarrow 10 = c$, $v(3) = 8 \Rightarrow 8 = 9a + 3b + c$, $v(6) = 16 \Rightarrow 16 = 36a + 6b + c$. Therefore, $9a + 3b = -2$, $36a + 6b = 6 \Rightarrow 6a + b = 1$. $b = 1 - 6a \Rightarrow 9a + 3(1 - 6a) = -2 \Rightarrow -9a + 3 = -2$
$\Rightarrow a = \dfrac{5}{9}$, $b = 1 - 6a = -\dfrac{7}{3}$. Therefore, $v(t) = \dfrac{5}{9}t^2 - \dfrac{7}{3}t + 10 \Rightarrow s(t) = \dfrac{5}{27}t^3 - \dfrac{7}{6}t^2 + 10t + c$.
But, $s(0) = 0 \Rightarrow c = 0$ so $\Rightarrow s(t) = \dfrac{5}{27}t^3 - \dfrac{7}{6}t^2 + 10t$ and $s(10) = 168.5$. So, after 10 seconds, the particle is 168.5 m from O.

6. (a) $s(t) = -12.5e^{-0.4t} + c$, $0 = -12.5 + c \Rightarrow c = 12.5 \Rightarrow s(t) = 12.5(1 - e^{-0.4t})$ and $s(8) = 12.5(1 - e^{-3.2}) = 11.99$.

(b) 12.5

(c) $a(t) = -2e^{-0.4t} \Rightarrow a(1) = -1.34$

Exercise 7.17

1. (a) 3.75 (b) 2 (c) $\dfrac{1}{6}$ (d) $\dfrac{8}{3}$ (e) 2.5 (f) 14.25

2. (a) $\displaystyle\int_{-1}^{1} e^x\,dx = \left[e^x\right]_{-1}^{1} = e - \dfrac{1}{e} = 2.35$ (b) $\displaystyle\int_{0}^{2} \dfrac{1}{2x+3}\,dx = \left[\dfrac{1}{2}\ln(2x+3)\right]_{0}^{2} = \dfrac{1}{2}\ln\dfrac{7}{3} = 0.424$

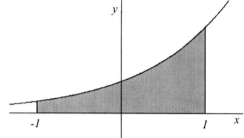
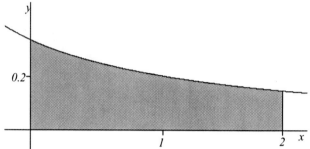

(c) $\displaystyle\int_{0}^{\pi/3} \cos x\,dx = [\sin x]_{0}^{\pi/3} = \dfrac{\sqrt{3}}{2} = 0.866$

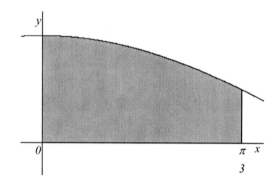

3. (a) $\displaystyle\int_{0}^{4} e^{-\frac{1}{2}t}\,dt = \left[-2e^{-\frac{1}{2}t}\right]_{0}^{4} = -2e^{-2} - (-2) = 2 - \dfrac{2}{e^2} = 2\left(1 - \dfrac{1}{e^2}\right)$

(b) $\displaystyle\int_{1}^{5} \dfrac{4}{(3x+1)}\,dx = \left[\dfrac{4}{3}\ln(3x+1)\right]_{1}^{5} = \dfrac{4}{3}(\ln 16 - \ln 4) = \dfrac{4}{3}\ln 4$

(c) $\displaystyle\int_{0}^{\pi/2} (1 + \sin 2\theta)\,d\theta = \left[\theta - \dfrac{1}{2}\cos 2\theta\right]_{0}^{\pi/2} = \left(\dfrac{\pi}{2} - \dfrac{1}{2}\cos\pi\right) - \left(0 - \dfrac{1}{2}\cos 0\right) = 1 + \dfrac{\pi}{2}$

4. The large sketches are not drawn, but the shape of the functions can be seen from the graphing calculator screen displays shown below.

(a) 1.7642   (b) 2.3963   (c) 19.6940

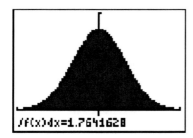 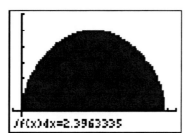 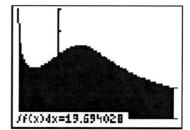

5. (a)

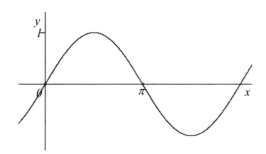

(b) (i) $[-\cos x]_0^\pi = 2$ (ii) $[-\cos x]_\pi^{2\pi} = -2$ (iii) $[-\cos x]_0^{2\pi} = 0$

(c) Area below the x-axis is measured negatively so that the net area between the curve and the x-axis between 0 and 2π is $2 + (-2) = 0$.

6. (a)   (b)

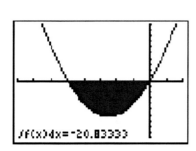

$$\text{Area} = \left|-\frac{125}{6}\right| = 20.8 \hspace{3cm} \text{Area} = |-1.6| = 1.6$$

(c)

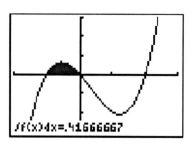 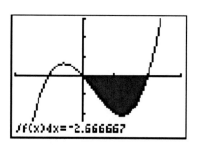

Area $= \dfrac{5}{12} \approx 0.417$ 	 Area $= \left|-\dfrac{8}{3}\right| \approx 2.67$

Therefore total area enclosed is $\dfrac{37}{12} \approx 3.08$

(d)

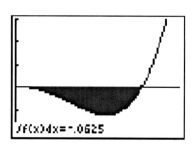

Area $= \left|-\dfrac{1}{16}\right| = 0.625$

(e)

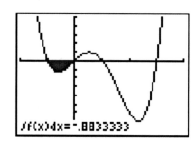 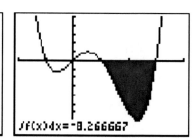

Therefore total area enclosed is $|-0.88333| + 0.61667 + |-8.26667| \approx 9.77$

(f) The curve intersects the x-axis in the following points: -0.56107, 0.59924, 1.34805

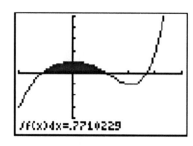 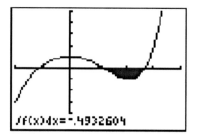

Therefore the total area enclosed is $0.77102 + |-0.49326| \approx 1.26$.

Exercise 7.18

1. (a) 162π (b) 111π (c) 18.6π (d) $\dfrac{\pi \ln 9}{4}$

2. It is advisable to use your graphing calculator. (a) 9.87 (b) 182 (c) 1.36 (d) 197

3. $V = \pi \int_0^3 (3x - x^2)^2 \, dx = \pi \int_0^3 (9x^2 - 6x^3 + x^4) \, dx = \pi \left[3x^3 - \dfrac{3}{2}x^4 + \dfrac{1}{5}x^5 \right]_0^3 = \pi \left(81 - 121.5 + \dfrac{243}{5} \right)$

   $= \dfrac{40.5}{5} \pi = 8.1\pi$

4. Let V be the volume of a hemisphere formed by rotating the curve $y = \sqrt{r^2 - x^2}$, $0 \le x \le r$ completely about the x-axis.

   Then $V = \pi \int_0^r (r^2 - x^2) \, dx = \pi \left[r^2 x - \dfrac{1}{3} x^3 \right]_0^r$

   $= \pi \left(r^3 - \dfrac{1}{3} r^3 \right) = \dfrac{2\pi r^3}{3}$.

   Therefore, the volume of a complete sphere is $2V = \dfrac{4\pi r^3}{3}$.

   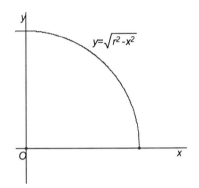

5. $V = \int_0^1 \pi (x^2 - x^4) \, dx = \pi \left[\dfrac{1}{3} x^3 - \dfrac{1}{5} x^5 \right]_0^1 = \pi \left(\dfrac{1}{3} - \dfrac{1}{5} \right) = \dfrac{2}{15} \pi$

Printed in the United States
77887LV00005B/126